Geologic Studies of Mercury by the U.S. Geological Survey

Edited by John E. Gray

U.S. Geological Survey Circular 1248

U.S. Department of the Interior
U.S. Geological Survey

U.S. Department of the Interior
Gale A. Norton, Secretary

U.S. Geological Survey
Charles G. Groat, Director

U.S. Geological Survey, Reston Virginia 2003

For sale by U.S. Geological Survey Information Services
Box 25286, Denver Federal Center
Denver, CO 80225

For more information about the USGS and its products:
Telephone: 1-888-ASK-USGS
World Wide Web: http://www.usgs.gov/

Foreword

It is the goal of the U. S. Geological Survey (USGS) to enhance and protect the quality of life by describing and understanding the Earth around us. This volume summaries selected geologic studies of mercury in the United States by the USGS. The six geologic studies report concentrations of mercury in coal, sediment, soil, water, air, and fish; and discuss how the USGS is evaluating some of these areas of potential environmental mercury contamination. This collection of studies represents only a small portion of the ongoing worldwide research on mercury, and is not intended to be a comprehensive reference on the geochemistry of mercury. Numerous other mercury studies are in progress in the USGS, in other governmental agencies, in industry, and at universities throughout the world. These studies address the USGS Mineral Resources Program goal, which is to gain an understanding of the influence of mineral deposits, mineralizing processes, and mineral-resource development on environmental integrity, ecosystems, public health, and geologic hazards.

Kathleen M. Johnson
Mineral Resources Program Coordinator

Contents

Figures

Tables

Introduction

By John E. Gray

Mercury in the Products We Use

A liquid at room temperature, mercury is a unique metal with unusual properties. Elemental mercury has long been used in thermometers because it responds to changes in temperature. In fact, mercury's many diverse properties have made it useful for many products. Mercury is a good metallic conductor with a low electrical resistivity; it has been used in electrical products including electrical wiring and switches, fluorescent lamps, mercury batteries, and thermostats (Eisler, 1987). Mercury also is used in navigational instruments to measure changes in temperature and pressure. In the medical field, mercury is used as a component in dental fillings and as a preservative in many pharmaceutical products. Mercury has been used in industrial and agricultural applications such as in the production of chlorine and caustic soda, in nuclear reactors, in plastic production, for the extraction of gold (amalgamation) during mining, as a fungicide in seeds and bulbs, and as an antifouling agent in paper, paper pulp, and paint (Sznopek and Goonan, 2000).

The Cycle of Mercury in the Environment

In addition to being contained in many products that people make and use, mercury is also present in a variety of forms in rocks, soil, water, coal, petroleum, and even air. Although the amount of mercury present in rocks and soils is generally minor, this mercury can make its way into air and water by evaporation and through natural weathering and erosion. Mercury is also added to the air in the form of gas and small particles that are naturally erupted from volcanoes throughout the world. (See Hinkley, this volume.) However, a significant amount of the mercury present in the Earth's atmosphere is from burning fossil fuels, mostly coal, but also petroleum products. Coal is a common fuel used in many power plants to generate electricity throughout the world. Although the amount of mercury in coal is minor, large amounts of coal are typically used in coal electrification plants. Thus, burning coal is the largest human-caused contributor of mercury to the atmosphere. (See Finkelman, this volume.) Some mercury is also emitted to the atmosphere by the combustion of petroleum products such as gasoline in automobile and airplane engines. Other sources of mercury to the land, water, and atmosphere include the mining of mercury and gold, forest fires, incineration of waste from treatment plants, emissions from landfills, evaporation from oceans, and industrial runoff, seepage, and discharges (fig. 1). No matter where on Earth it originates, airborne mercury gas mixes in the atmosphere, can travel many thousands of miles, and is eventually redistributed around the world. Estimates suggest that the total contribution of mercury to the atmosphere from all sources worldwide is as much as 6,000 t (metric tons)/year (table 1). Some of the mercury in the atmosphere eventually deposits (most commonly in rain) on the Earth's surface in soils, sediments, oceans, rivers, and lakes (Mason and others, 1994). Although the concentration of mercury in lake and ocean water is low, these water bodies are large reservoirs of mercury because they have such a large volume of water. Mercury also evaporates from lakes and oceans, again contributing mercury to the atmosphere, completing the cycle.

Where Mercury Comes From

Mercury has been mined for more than 2,000 years and most of the mercury used historically by man has been produced through the mining of ore. Although mercury constitutes less than 1/100,000,000 of one percent of the Earth, and like many minerals and elements found in nature, mercury can be highly enriched in certain rocks called ore deposits. The most common mineral containing mercury in ore deposits is cinnabar, or mercury sulfide (HgS), but naturally occurring elemental mercury, or quicksilver (Hg°), is also found in some mercury deposits. Both cinnabar and elemental mercury are distinctive, making their identification relatively easy. Elemental mercury is a silver-colored liquid at room temperature (fig. 2); cinnabar is a distinctive red mineral (fig. 3). Roasting the ore in a furnace (fig. 4) easily converts cinnabar to elemental mercury; this ease of conversion is another reason why mercury has been mined for such a long time. Elemental mercury is the final product obtained through mining of cinnabar. The international unit of measurement of elemental mercury is a flask, which weighs about 34.5 kg or 76 pounds.

Historically, the largest mercury mines have been those in Spain, Italy, Slovenia, Peru, China, the former U.S.S.R., Algeria, Mexico, Turkey, and the State of California (fig. 5), but many other mercury mines are scattered throughout the world. Most mercury mines are presently closed owing to low demand and low prices for mercury worldwide, primarily as a result of environmental and health concerns surrounding mercury. Furthermore, considerable amounts of mercury-containing products are being recycled, especially in the United States, which also reduces the demand for mercury mining (Sznopek and Goonan, 2000). Although few mercury mines in the world are presently operating, closed and inactive mercury mines are sites of some of the highest mercury concentrations on Earth. At these mercury mines, mine wastes contain considerable cinnabar, elemental mercury, and other mercury compounds that are continually lost to surrounding environments through erosion, leaching, and evaporation.

Another significant mining use of mercury worldwide is the amalgamation of gold by mercury, a technique used for the extraction of precious metals in many mines. Although this practice is not generally used in the United States, it is still used in many developing countries. As a result of amalgamation practices, significant liquid mercury is lost to streams and rivers surrounding many gold mining areas throughout the world. In some of these areas, liquid mercury that was used decades ago remains in these rivers as a potential environmental problem. (See example of the Carson River, Nevada; Lawrence, this volume.)

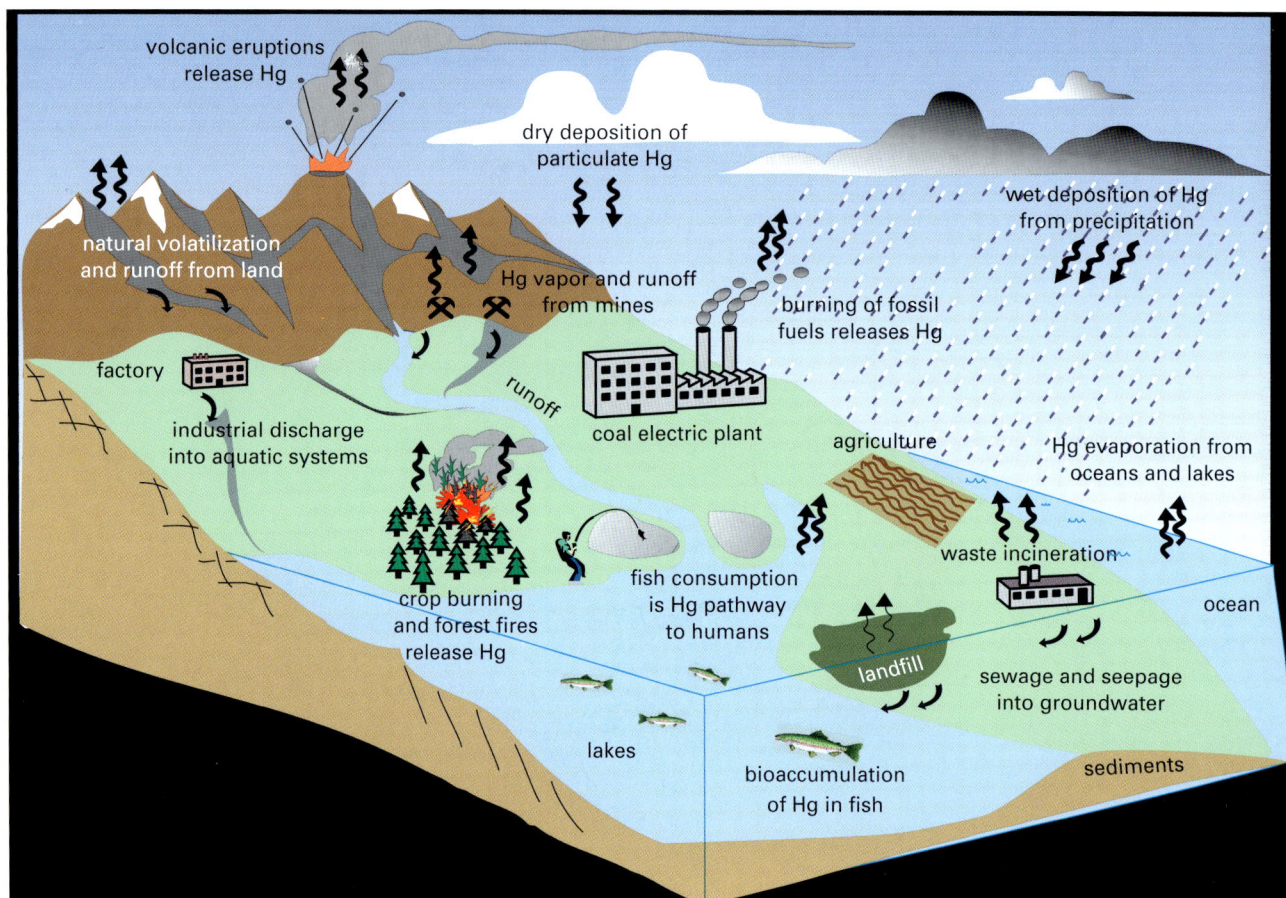

Figure 1. Schematic diagram of mercury cycle showing important contributions of mercury to the environment from land, water, air, and anthropogenic sources.

Table 1. Concentrations of mercury in the atmosphere and contributions of mercury to the atmosphere from natural and anthropogenic sources.

[Mercury concentrations given (ng/m^3) are those in air above the sources listed, which are provided to give a relative comparison of various sources contributing mercury to the atmosphere. Estimated mercury contributed to the atmosphere represents that from all like sources combined throughout the world, for example, all oceans worldwide are estimated to contribute as much as 2,000 t of mercury per year to the atmosphere. ≈, approximated; ng/ m^3, nanograms/cubic meter; kg/yr, kilograms/year; t/yr, metric tons/year]

Source	Hg concentration or emission*	Estimated Hg contributed to atmosphere worldwide	References
Atmosphere	1 – 2 ng/m^3	4,400 – 6,000 t/yr	Fitzgerald (1986); Porcella (1994); Lamborg and others (2002).
Volcanoes	28 – 1,400 ng/m^3	60 t/yr [#]	Fitzgerald (1986); Varekamp and Buseck (1986); Ferrara and others (1994).
Land	1 – 6 ng/m^3	≈ 1,000 t/yr	Varekamp and Buseck (1986); Gustin and others (1994); Mason and others (1994).
Mines	2 – 5,000 ng/m^3	≈10 – 100 t/yr	Ferrara and others (1991; 1998) Gustin and others (1994; 1996; 2000).
Oceans	1 – 3 ng/m^3	800 – 2,000 t/yr	Mason and others (1994); Lamborg and others (2002).
Anthropogenic	~10->900 kg/yr *	2,000 – 2,600 t/yr	Mason and others (1994); Environmental Protection Agency (2000); Lamborg and others (2002).

[#] Mercury emissions from some explosively erupting volcanoes indicate that this source of mercury to the atmosphere could be several times higher than shown here (Varekamp and Buseck, 1986).

*Range of mercury emissions in 1999 from numerous power plants in the U.S.A. as reported to the EPA (Environmental Protection Agency, 2000).

Figure 2. Gold pan with elemental mercury obtained from mercury-contaminated sediments. Such elemental mercury presently remains in sediments and abandoned sluice boxes at sites of historic gold mining in the California Sierra Nevada.

Figure 3. A sample of cinnabar (red mineral), the most common mercury ore in mercury mines worldwide.

The Chemistry of Mercury

The properties and behavior of mercury depend on its oxidation state. Mercury in nature is found in three oxidation states: metallic or elemental mercury (Hg°), *mercurous ion (Hg_2^{2+}), and mercuric ion (Hg^{2+})*. All three forms of mercury present some degree of hazard to life forms—including humans—but mercury compounds containing the mercuric ion are the most toxic, especially organic-mercury compounds (World Health Organization, 1976). All forms of mercury can be converted to these toxic organic compounds, and thus, all mercury compounds are considered potentially dangerous. Mercury in water, soil, sediments, biota, and rocks and minerals is mostly in the form of inorganic ionic compounds and organic compounds (fig. 6). Most of the mercury in air is Hg°, and oxidized forms (for example, Hg^{2+}) generally constitute less than 2 percent of the mercury in air (Fitzgerald, 1989).

Any form of mercury that makes its way into an aquatic system has the potential to be converted into organic mercury, of which methylmercury (CH_3Hg^+) is generally the most toxic. The process of methylmercury formation (mercury methylation) is complex and results from chemical and biological activity; mercury methylation is dependent on pH, temperature, oxidation-reduction potential, the amount of organic matter, and other chemical factors (Ullrich and others, 2001). Bacterial conversion of inorganic mercury to methylmercury is the dominant methylation process typically in the sediment column in aquatic environments (Compeau and Bartha, 1985). An important mechanism of methylmercury formation around mines is the oxidation of Hg° to Hg^{2+}, and the subsequent microbial formation of methylmercury. Methylmercury compounds are highly stable, are soluble in water and in the fats of organisms, and have the ability to penetrate membranes of living organisms. Once mercury is converted to methylmercury, biota in aquatic ecosystems rapidly absorb the mercury, and as a result, mercury tends to concentrate in tissues of fish and other aquatic organisms (bioaccumulation). Mercury also biomagnifies in the food chain, and it generally increases with increasing position in the food chain when environments are exposed to mercury (fig. 7). The most common pathway of mercury to humans and other higher order wildlife is through diet, primarily through consumption of fish and seafood products (Ullrich and others, 2001). Animals and humans that consume large quantities of fish are at the highest risk of mercury contamination because the form of mercury in fish is mostly (generally > 90 percent) highly toxic methylmercury (National Academy of Sciences, 1978; Clarkson, 1990; U.S. Environmental Protection Agency, 1997). Methylmercury is more thoroughly absorbed in the human gastrointestinal system (about 95 percent absorption) compared to other forms of mercury such as elemental mercury (less than 10 percent absorption) (U.S. Environmental Protection Agency, 1997). Although humans take in most mercury through food sources, exposure breathing elemental mercury vapor is also possible, but more rare. For these reasons, scientists often measure the concentration of mercury and methylmercury to evaluate mercury contamination in specific areas.

Figure 4.   Inactive mercury mine in Nevada. In the rotary furnace, mercury ore was burned producing mercury gas that was cooled, condensed as elemental mercury, and collected.

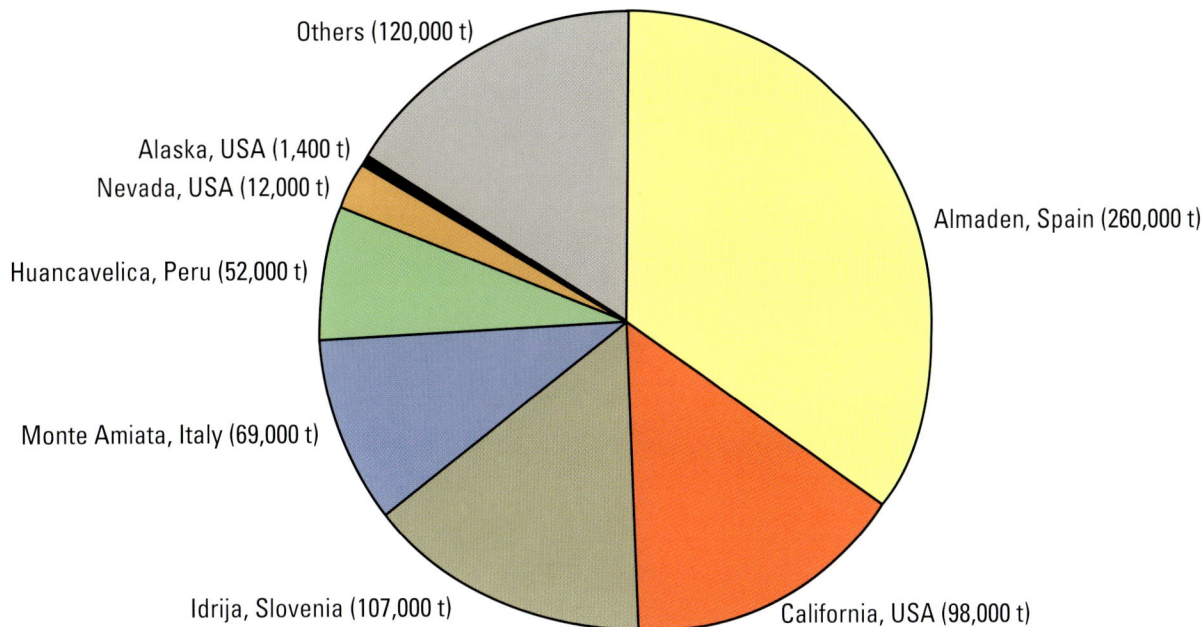

Figure 5. Production of mercury (metric tons) from some mercury mines found throughout the world. Production from mines in China, the former U.S.S.R., Algeria, Mexico, and Turkey are grouped as "others."

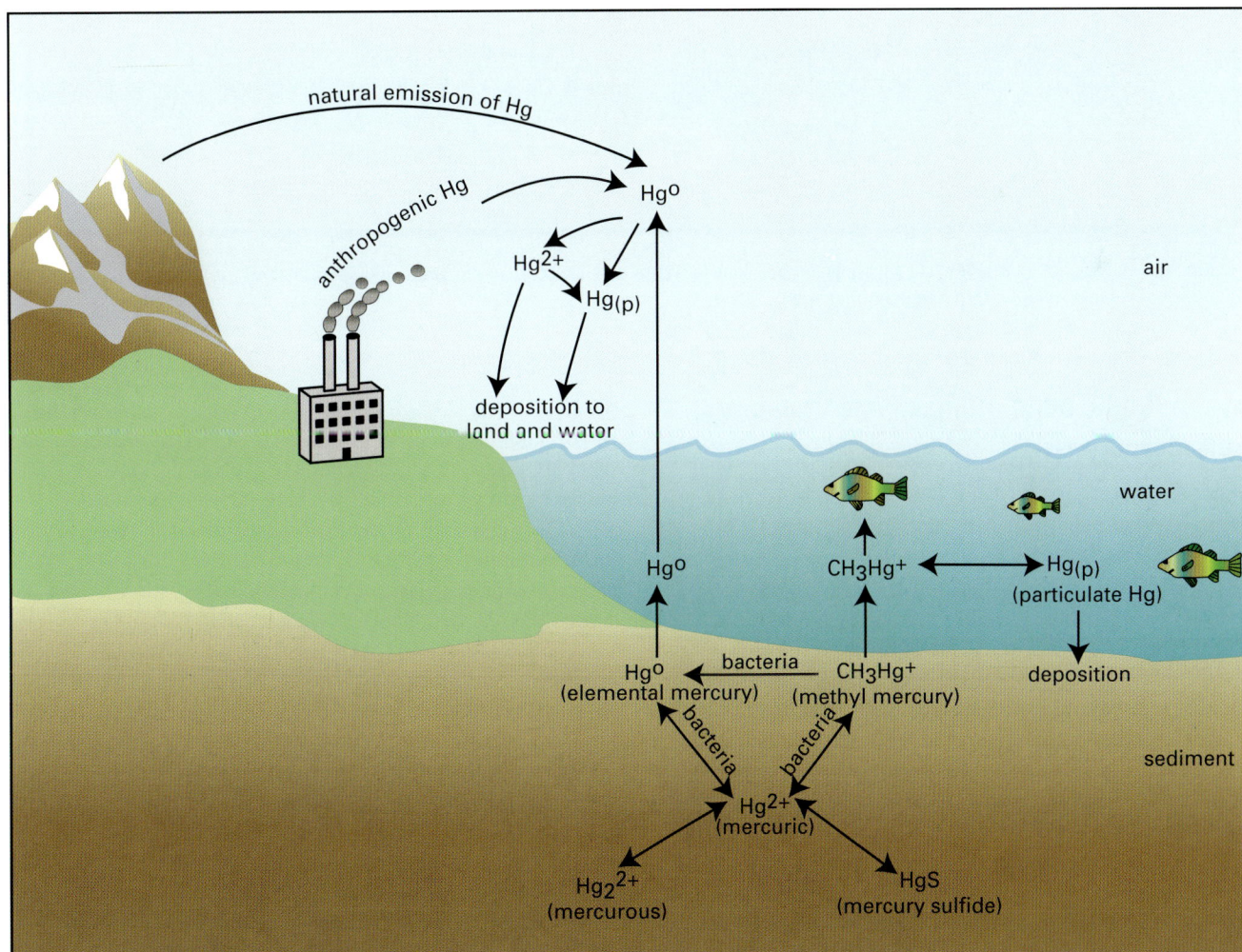

Figure 6. Major species and transformations of mercury in the environment (modified from Wood, 1974; Mason and others, 1994). Conversion to methylmercury is most important because it is bioavailable and is transferred to water and biota.

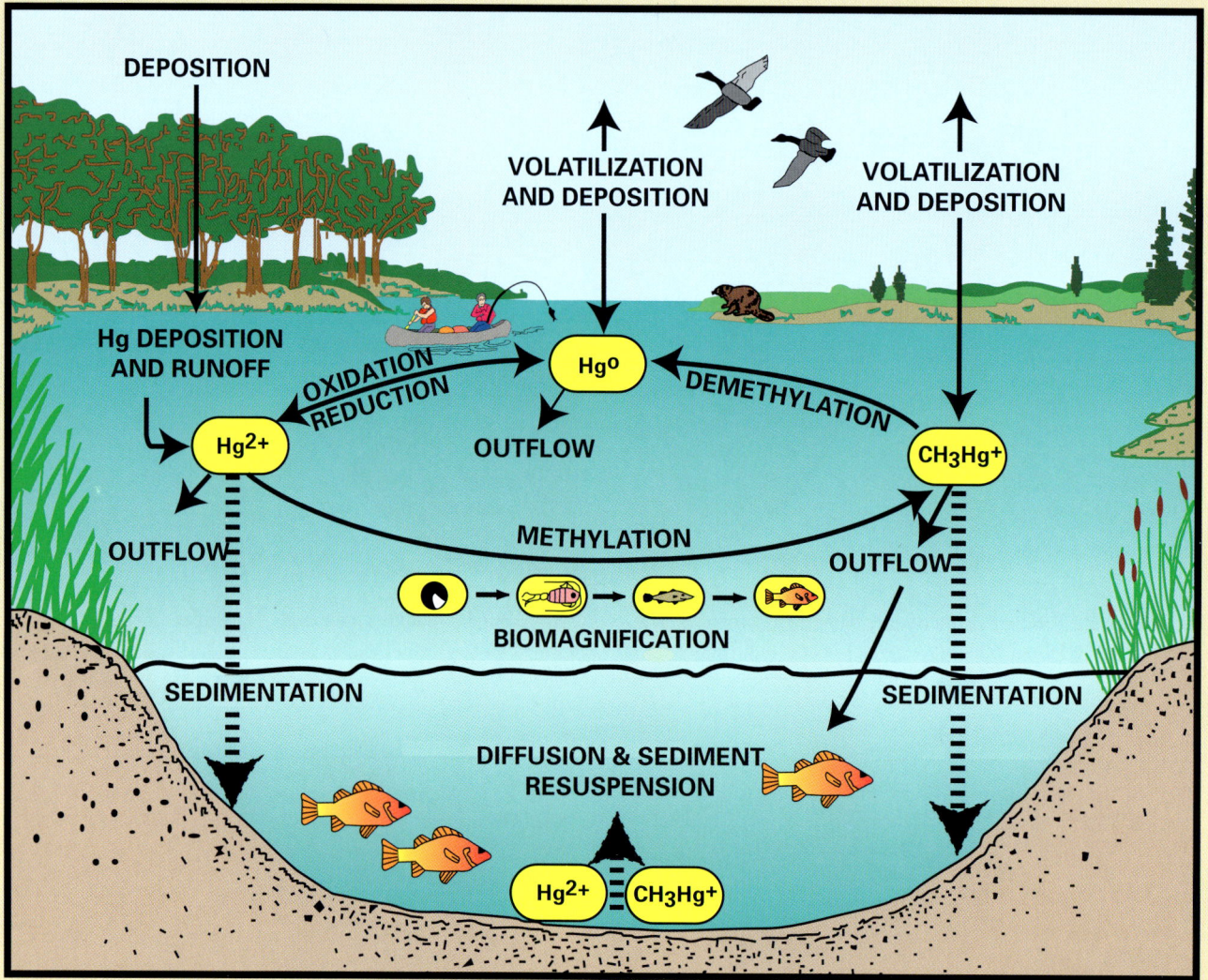

Figure 7. The aquatic mercury cycle showing important mercury species, mercury methylation and demethylation, and biomagnification of mercury in biota (modified from Hudson and others, 1994).

Why the Public and Scientists Are Concerned About Mercury

Mercury is a heavy metal of environmental concern because elevated concentrations can be toxic to all living organisms. Mercury has no known metabolic function in animals and is not easily eliminated by organisms, including humans. High concentrations of mercury in humans adversely affect the central nervous system, especially the sensory, visual, and auditory parts that effect coordination (Fitzgerald and Clarkson, 1991). In extreme cases, mercury poisoning can lead to death (National Academy of Sciences; 1978; Eisler, 1987; U.S. Environmental Protection Agency, 1997). For all organisms, the early stages of development (especially embryos) are the most sensitive to mercury (Clarkson 1990; U.S. Environmental Protection Agency, 1997).

For several decades, scientists and governments have been aware of the toxic effects of mercury on humans and other organisms. Since the 1950s, numerous cases of mercury poisoning to humans and wildlife have occurred in Japan, Iraq, Scandinavia, Europe, the United States, Canada, the Philippines, and in

the Amazon River region of South America (Eisler, 1987; U.S. Environmental Protection Agency, 1997). These cases resulted from high concentrations of mercury in aquatic systems contaminated by industrial discharges, in agricultural products, by atmospheric deposition, by exposure to elemental mercury, and through mining activities. In North America, numerous lakes and reservoirs are known to contain fish that have mercury concentrations above the level considered safe for human consumption (Krabbenhoft and Rickert, 1995). As of December 2000, more than 2,200 water bodies in 41 States in the United States have advisories for high concentrations of mercury in fish, and State and Tribal governments have advised citizens against eating these fish (U.S. Environmental Protection Agency, 2001). In some of these instances, such as in California, mercury contamination is clearly related to past mining activities. (See Hunerlach and Alpers, this volume.) However, for most of these aquatic advisories, the source of the mercury is not related to any obvious mining or industrial discharges. In these cases, atmospheric mercury deposition is more probable (Fitzgerald and others, 1998; Krabbenhoft and Wiener, 1999), but these aquatic systems need additional study.

Why the USGS is Studying Mercury

As a result of its wide use, mercury is common in the environment in which we live. Because of its toxic nature, numerous studies of environmental mercury contamination are ongoing in the United States and throughout the world. In 1997, the U.S. Environmental Protection Agency (EPA) completed a report that was submitted to the U.S. Congress as required under the 1990 Clean Air Act (U.S. Environmental Protection Agency, 1997). In this study, the EPA evaluated many aspects of mercury emissions in the U.S., the health and environmental implications of those emissions, and the availability and cost of emission control technologies. The EPA report also recommended areas for future research to aid in the understanding of sources, transport, and health effects related to mercury in the air, water, and through fish consumption. In addition, the EPA report contained recommendations concerning possible monitoring, control, remediation, and regulation of environmental mercury problems in the United States. Although the USGS is generally not involved in remediation or regulatory practices, the USGS contributes, through various monitoring and research studies, to the overall understanding of the geochemistry of mercury and how it enters and affects the environment.

References Cited

Clarkson, T.W., 1990, Human health risks from methylmercury in fish: Environmental Toxicology and Chemistry, v. 9, p. 957–961.

Compeau, C.C., and Bartha, R.A., 1985, Sulfate reducing bacteria—Principal methylators of mercury in anoxic estuarine sediment: Applied Environmental Microbiology, v. 50, p. 498–502.

Eisler, R., 1987, Mercury hazards to fish, wildlife, and invertebrates—A synoptic review: U.S. Fish and Wildlife Service Biological Report 85 (1.10), 90 p.

Ferrara, R., Maserti, B.E., Andersson, M., Edner, H., Ragnarson, P., Svanberg, S., and Hernandez, A., 1998, Atmospheric mercury concentrations and fluxes in the Almaden District (Spain): Atmospheric Environment, v. 32, p. 3897–3904.

Ferrara, R., Maserti, B.E., Breder, R., 1991, Mercury in abiotic and biotic compartments of an area affected by a geochemical anomaly (Mt. Amiata, Italy): Water, Air, and Soil Pollution, v. 56, p. 219–233.

Ferrara, R., Maserti, B.E., De Liso, A., Cioni, R., Raco, B., Taddeucci, G., Edner, H., Ragnarson, P., Svanberg, S., and Wallinder, E., 1994, Atmospheric mercury emission at Solfatara Volcano (Pozzuoli, Phlegraean fields, Italy): Chemosphere, v. 29, no. 7, p. 1421–1428.

Fitzgerald, W.F., 1986, Cycling of mercury between the atmosphere and oceans, in Buat-Menard, P., ed., The role of air-sea exchange in geochemical cycling: NATO, Advanced Science Institutes Series, Dordrecht, Reidel Publishing Co., p. 363–408.

Fitzgerald, W.F., 1989, Atmospheric and oceanic cycling of mercury, in Ripley, J.P., and Chester, R., eds., Chemical oceanography: New York, Academic Press, p. 151–186.

Fitzgerald, W.F., and Clarkson, T.W., 1991, Mercury and monomethylmercury—Present and future concerns: Environmental Health Perspectives, v. 96, p. 159–166.

Fitzgerald, W.F., Engstrom, D.R., Mason, R.P., and Nater, E.A., 1998, The case for atmospheric mercury contamination in remote areas: Environmental Science and Technology, v. 32, no. 1, p. 1–7.

Gustin, M.S., Lindberg, S.E., Austin, K., Coolbaugh, M., Vette, A., and Zhang, H., 2000, Assessing the contribution of natural sources to regional atmospheric mercury budgets: Science of the Total Environment, v. 259, p. 61–71.

Gustin, M.S., Taylor, G.E., Jr., and Leonard, T.L., 1994, High levels of mercury contamination in multiple media of the Carson River drainage basin of Nevada—Implications for risk assessment: Environmental Health Perspectives, v. 102, no. 9, p. 772–778.

Gustin, M.S., Taylor, G.E., Jr., Leonard, T.L., and Keislar, R.E., 1996, Atmospheric mercury concentrations associated with geologically and anthropogenically enriched sites in central western Nevada: Environmental Science and Technology, v. 30, no. 8, p. 2572–2579.

Hudson, R.J.M., Gherini, S.A., Watras, C.J., and Porcella, D.B., 1994, Modeling the biogeochemical cycle of mercury in lakes—The mercury cycling model (MCM) and its application to the MTL study in lakes, in Watras, C.J., and Huckabee, J.W., eds., Mercury pollution, integration and synthesis: Boca Raton, Fla., CRC Press, p. 473–523.

Krabbenhoft, D.P., and Rickert, D.A., 1995, Mercury contamination of aquatic ecosystems: U.S. Geological Survey Fact Sheet FS-216-95.

Krabbenhoft, D.P., and Wiener, J.G., 1999, Mercury contamination—A nationwide threat to our aquatic resources, and a proposed research agenda for the U.S. Geological Survey, in Morganwalp, D.W., and Buxton, H.T., eds., U.S. Geological Survey, Toxic Substances Hydrology Program, Water-Resources Investigations Report 99-4018B, p. 171–178.

Lamborg, C.H., Fitzgerald, W.F., O'Donnell, J., and Torgersen, T., 2002, A non-steady-state compartmental model of global-scale mercury biogeochemistry with interhemispheric atmospheric gradients: Geochimica et Cosmochimica Acta, v. 66, p. 1105–1118.

Mason, R.P., Fitzgerald, W.F., and Morel, F.M.M., 1994, The biogeochemistry of elemental mercury—Anthropogenic influences: Geochimica et Cosmochimica Acta, v. 58, p. 3191–3198.

National Academy of Sciences, 1978, An assessment of mercury in the environment: Washington, D.C., National Academy of Sciences, National Research Council, 185 p.

Porcella, D.B., 1994, Mercury in the environment—Biochemistry, in Watras, C.J., and Huckabee, J.W., eds., Mercury pollution, integration and synthesis: Boca Raton, Fla., CRC Press, p. 3–19.

Sznopek, J.L., and Goonan, T.G., 2000, The materials flow of mercury in the economies of the United States and the world: U.S. Geological Survey Circular 1197, 28 p.

Ullrich, S.M., Tanton, T.W., and Abdrashitova, S.A., 2001, Mercury in the aquatic environment—A review of factors affecting methylation: Critical Reviews in Environmental Science and Technology, v. 31 (3), p. 241–293.

U.S. Environmental Protection Agency, 1997, Mercury study report to Congress, Volumes I–VIII: U.S. Environmental Protection Agency Report EPA-452/R-97-003.

U.S. Environmental Protection Agency, 2000, Emissions and Generation Resource Integrated Database (E-GRID): http://www.epa.gov/airmarkets/egrid/, accessed 12/10/02.

U.S. Environmental Protection Agency, 2001, Mercury update—Impact on fish advisories: U.S. Environmental Protection Agency Fact Sheet EPA-823-F-01-011.

Varekamp, J.C., and Buseck, P.R., 1986, Global mercury flux from volcanic and geothermal sources: Applied Geochemistry, v. 1, p. 65–73.

Wood, J.M., 1974, Biological cycles for toxic elements in the environment: Science, v. 183, p. 1049–1052.

World Health Organization, 1976, Environmental Health Criteria no. 1—Mercury: Geneva, Switzerland, World Health Organization.

Mercury in Coal and Mercury Emissions from Coal Combustion

By Robert B. Finkelman

Abstract

Mercury emissions from coal-fired electric generating utilities are a major uncontrolled source of mercury in the environment. This mercury may be contributing to serious health problems in segments of our society. The USGS is compiling information on mercury in coal that may be useful in developing strategies for reducing mercury emissions from coal use. The USGS coal-quality database contains information on mercury concentrations in more than 7,000 coal samples. Detailed geochemical analysis has helped to determine that mercury in coal is commonly associated with pyrite, but other modes of occurrence may be locally important. Physical coal cleaning removes, on average, 37 percent of the mercury. Characterization of feed coal and its combustion byproducts is helping to further understand the behavior of mercury in utility boilers.

Introduction

The concentration of mercury in coal has been of concern since the passage of the 1990 Clean Air Act Amendment (Toole-O'Neil and others, 1999). In 1994, the U.S. Environmental Protection Agency (EPA) estimated that about 50 t of mercury is emitted each year from coal-burning power plants in the United States, with lesser amounts coming from oil- and gas-burning units. In February 1998, the EPA issued a report citing mercury emissions from electric utilities as the largest uncontrolled source of mercury to the atmosphere (U.S. Environmental Protection Agency, 1997). The EPA estimates emissions from coal-fired utilities (fig. 8) may exceed 25 percent of the total airborne emissions of mercury (natural plus anthropogenic) in the United States (table 1). The EPA suggested that utility mercury emissions are of sufficient potential concern for public health to merit further research and monitoring (U.S. Environmental Protection Agency, 1997). For the past 20 years, the USGS has been conducting research on the distribution and concentration of mercury in coal in the United States. More recently, the USGS has undertaken research to understand the forms of mercury in coal and its behavior during coal cleaning and combustion (Toole-O'Neil and others, 1999).

Coal-Quality Database on Mercury

The USGS has compiled a coal-quality database containing information on the concentration of mercury in more than 7,000 in-ground coal samples. The average concentration of total

Figure 8. Four Corners coal-fired electrification power plant near Farmington, N. Mex. Emissions from such power plants that use coal for fuel are under study as sites of potential mercury contamination to local and regional environments.

mercury in coal is about 0.2 µg/g (micrograms/gram); values exceeding 1 µg/g are rare. On an equal energy basis, the highest mercury concentrations are found in the Gulf Coast lignites (36 lb of Hg/10^{12} Btu), and the Hams Fork region coal (Wyoming) has the lowest value (4.8 lb of Hg/10^{12}Btu). Mean concentrations for total mercury in coal for the major coal basins in the United States are shown in figure 9. The data for individual samples can be found at: http://energy.er.usgs.gov/products/databases/coalqual/intro.htm.

The USGS is also developing a database that will contain information on the mercury content of coals being mined and burned in other major coal-producing countries. The data should help to establish worldwide contributions of mercury as a result of emissions from coal combustion—information that is presently not well known

Forms of Mercury in Coal

As a result of the generally low concentration of total mercury in coal, and the high volatility of mercury, it is particularly difficult to determine the form(s) of mercury in coal. Recent research indicates that most of the mercury in coal is associated generally with secondary, arsenic-bearing pyrite (Finkelman, 1981; Toole-O'Neil and others, 1999). The mercury was deposited with the pyrite in cleats and fractures when hydrothermal solutions percolated through coal (Toole-O'Neil and others, 1999). Other forms of mercury that have been reported are organically bound mercury, elemental mercury, and mercury sulfides and selenides (fig. 10); (Finkelman, 1981). Mercury selenides may be the primary form of mercury in coal samples with little pyrite.

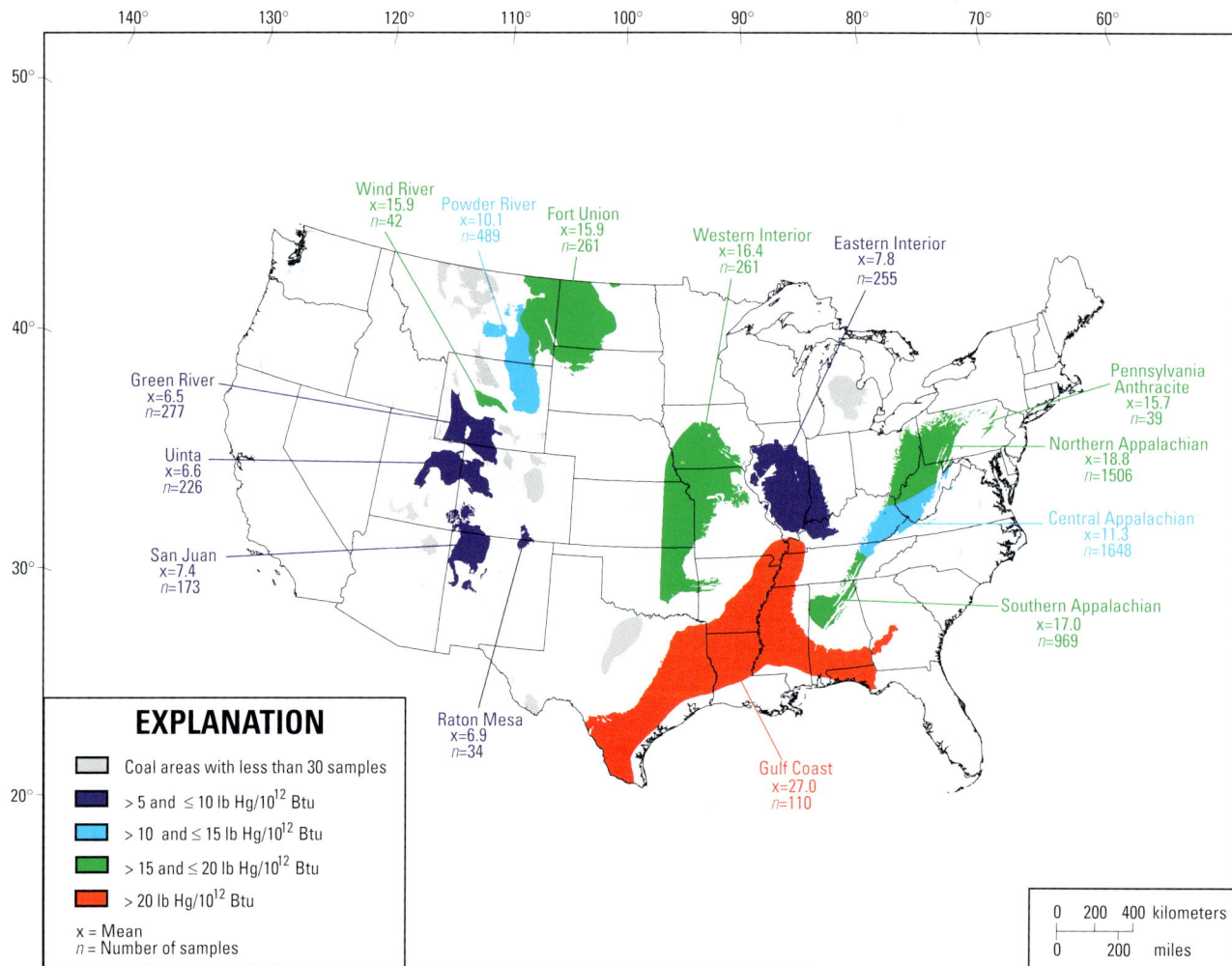

Figure 9.   Distribution of mercury in coal fields in the United States. Data are given on an equal energy basis, in which mercury concentration is converted to weight percent and divided by Btu.

Reducing Mercury Emissions from Power Plants

Because mercury is often found in fracture-filling pyrite, conventional coal cleaning procedures are generally effective in reducing mercury levels in the coals being burned. Conventional coal cleaning typically uses physical methods based on density differences to separate coal and minerals such as pyrite. The USGS is researching ways to assess the ability to remove mercury from coal by conventional physical coal cleaning techniques. The results of these studies indicate that an average of 37 percent of the mercury is removed by commercial coal cleaning procedures (Toole-O'Neil and others, 1999).

In addition to coal cleaning, several other methods exist for reducing mercury emissions. These include *fuel switching*—switching to oil or gas or using coal having lower mercury contents; *selective mining*—disposing of or simply not mining parts of the coal bed or deposit that have high mercury contents; *modifying combustion conditions*—such as using fluidized-bed combustion[1]; *post-combustion pollution control*—use of

electrostatic precipitators or baghouses (fabric filter traps) to capture particulates, or flue-gas desulfurization systems to remove pollutants from the gaseous effluents. New pollution control systems, such as activated carbon injection specifically designed for mercury capture, are presently being developed and tested. Research scientists at the USGS are studying the distribution of mercury and the forms of mercury in coal that will be helpful for evaluating mercury pollution from coal and assessing ways to reduce such pollution through fuel switching, selective mining, and physical coal cleaning.

Continuing USGS Research on Coal Combustion

As discussed above, postcombustion pollution-control systems can remove some of the mercury volatilized by coal combustion. Chu and Porcella (1995) indicated that electrostatic precipitators and fabric filters that trap fly ash remove about 30 percent of the mercury. The effectiveness of mercury removal by wet flue-gas-desulfurization systems varies widely, but on average about 45 percent, and as much as 90 percent, of the total mercury can be eliminated. The USGS had a project to determine the mercury content (as well as the concentrations of about 45 other elements) in feed coal, fly ash, bottom ash, and

---

[1] This technique uses pulverized coal that is suspended by upward-flowing air in a "bed" of particles, commonly limestone. This process operates at lower temperatures than conventional power generators, and the limestone captures pollutants such as mercury.

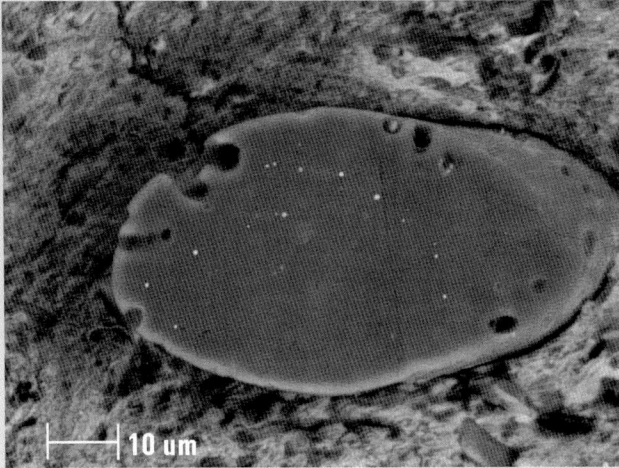

Figure 10. Scanning electron micrograph of mercury selenide (HgSe) crystals (white spots) in lignite coal from California. Mercury selenide is a highly toxic mercury compound and is a rare example of a mercury mineral in coal. Scale bar is 10 micrometers.

monitoring period. Therefore, few high-quality analyses of representative samples may adequately describe the Hg content of some feed coals.

flue-gas- desulfurization products (Breit and others, 1996). In this study, long-term (monthly sampling for 2 years) monitoring of coal samples from a power plant with units burning high- and low-sulfur coal indicated little variation in mercury content in either the feed coal or the coal-combustion products during the

References Cited

Breit, G., Eble, C., Finkelman, R., Affolter, R., Belkin, H., Brownfield, M., Cathcart, J., Hower, J., Leventhal, J., McGee, J. Palmer, C., Reynolds, R., Rice, C., and Zielinski, R., 1996, Systematic investigation of the compositional variations in solid waste products of coal combustion: 13th Annual International Pittsburgh Coal Conference, p. 1356–1361.

Chu, P., and Porcella, D.B., 1995, Mercury stack emissions from U.S. electric utility power plants, *in* Porcella, D.B., Huckabee, J.W., and Wheatley, B., eds., Mercury as a global pollutant: Water, Air, and Soil Pollution, v. 80, p. 135–144.

Finkelman, R.B., 1981, Modes of occurrence of trace elements in coal: U.S. Geological Survey Open-File Report 81-99, 314 p.

Toole-O'Neil, B., Tewalt, S.J., Finkelman, R.B., and Akers, D.J., 1999, Mercury concentration in coal—Unraveling the puzzle: Fuel, v. 78, p. 47–54.

U.S. Environmental Protection Agency, 1997, Mercury study report to Congress, Volume I, Executive summary: U.S. Environmental Protection Agency Report EPA-452/R-97-003.

Mercury Mine Studies

Environmental Impact of Mercury Mines in the Coast Ranges, California

By James J. Rytuba

Abstract

The California Coast Ranges mercury mineral belt has been the largest producer of mercury in North America. Two types of mercury deposits are found in the mineral belt: hot spring and silica-carbonate. The primary ore mineral in both mercury deposit types is cinnabar, but small amounts of elemental mercury are present. Mercury mines recovered mercury by heating ores in a furnace or retort to a temperature above the stability of cinnabar in order to release mercury vapor. During roasting of mercury ores, new mercury compounds formed. The speciation of mercury in mine tailings indicates that metacinnabar, corderoite, schuetteite, and mercury chlorides formed during the processing of ores.

Mine drainage is associated with many of the mercury deposits, and the geology and geochemistry of the deposit is important in determining the pH and composition of mine drainage from the two types of mercury deposits. The presence of carbonate minerals and serpentinite associated with silica-carbonate deposits serve to mitigate the acidity of mine drainage except where high iron-sulfide content occurs and resultant acidity is as low as pH 2. High concentrations of total mercury (up to 450,000 ng/L) and methylmercury (up to 70 ng/L) were found in mine-drainage waters from both deposit types. Mercury and methylmercury from mine drainage is adsorbed onto iron-rich precipitates and is seasonally flushed in streams during periods of high water flow.

Introduction

Mercury has been mined in North America since the early 1800s with over 3,800,000 flasks (about 130,000 t) of mercury being produced from several mercury mineral belts (fig. 11). The California Coast Ranges mineral belt has been the largest producer of mercury. It contains more than 50 mines that have produced more than 1,000 flasks each—including New Almaden (fig. 12), the largest mercury mine in North America, which produced about 2,800,000 flasks. Much of the elemental mercury produced in North America was used in the recovery of gold from placer and hard-rock mines, using the mercury amalgamation process. Because of environmental concerns and the consequent low price of mercury, large-scale mercury mining ceased by about 1990 in North America. In the United States, mercury is now produced only as a byproduct from presently operating gold mines where environmental regulations require its recovery and from the reprocessing of precious-metal-mine tailings and gold-placer sediments.

Many of the mercury mines in the California Coast Ranges pose an environmental concern because of the presence of mine waste rock that contributes mercury-rich sediment to nearby watersheds. At some of the mines, acidic drainage adversely affects the water quality of surrounding streams. The release of mercury in mine drainage is a significant source of mercury to watersheds, where it may bioaccumulate in biota, including fish.

Mineralogy of the Coast Ranges Mercury Deposits

The California mercury mineral belt extends for 400 km in the southern and central Coast Ranges and contains two distinct types of mercury deposits—silica-carbonate and hot-spring deposits (fig. 12); (Rytuba, 1996). The primary ore mineral in both these deposits is cinnabar, but elemental mercury is also found in small amounts (table 2). In some of the silica-carbonate deposits, metacinnabar, the high temperature polymorph of cinnabar, is an important ore mineral. Mercury chloride and mercury sulfate minerals are rare, but in a few deposits they constitute the primary ore minerals. In hot-spring deposits, pyrite and native sulfur are found in only small amounts. In silica-carbonate deposits, iron-sulfide minerals such as pyrite and marcasite are common. In some silica-carbonate deposits, iron sulfides constitute as much as 50 percent of the ore minerals, but iron sulfides are generally minor in hot-spring deposits. Iron sulfides tend to be environmentally adverse because they generate acidic water upon weathering.

Mercury Compounds in Mine Wastes

Mercury mines in the Coast Ranges were typically small, affecting areas of a square kilometer or less. The mercury ores were mined and processed at the same site, and only rarely were ores transported to a central processing facility. The primary recovery method consisted of roasting mercury ore in a furnace or retort to a temperature above that for the stability of cinnabar in order to release mercury vapor and sulfur. The mercury vapor was then cooled in a condenser system, and elemental mercury was recovered in a water-cooled trough at the base of the condensing columns. Inefficiencies in the roasting process produced mercury vapor and mercury-rich particulates that were released to the atmosphere and deposited downwind from the furnace site. Soot that accumulated in the condensing columns was periodically removed and reprocessed in a retort to recover any remaining mercury. Condenser soot has the consistency of

Figure 11.   Mercury mineral belts in North America. Mercury mines shown have significant mercury production (> 1,000 flasks or 34 t), and mercury occurrences have little or no production. Modified from Rytuba (2003).

ash and can be readily redistributed by wind. Discarded soot is an environmental concern because it may contain as much as several weight percent mercury, primarily as elemental mercury and soluble mercury sulfates and chlorides.

The process of heating ore to vaporize mercury from ore is a type of calcination process, and the resultant mercury mine wastes are termed calcines. These calcines have a characteristic red color that results from the oxidation of iron sulfides during ore roasting and the presence of fine-grained cinnabar (fig. 13). Lime (CaO) was also added to the mercury ore to remove sulfur. Mercury mine-waste calcines were typically discarded adjacent to the furnace site or into nearby stream channels. Flood events periodically transported the calcines downstream, thus continually providing space for disposal of additional mine wastes. As a result, calcines are typically found in stream channels and over-bank material for several kilometers downstream from mines. Mine wastes were also used for road construction adjacent to mines as another method of discarding the wastes.

In California, the concentration of total mercury in calcines typically ranges from 10 to 1,500 µg/g depending on the efficiency of the roasting process. In addition to the concentration

of total mercury in calcines, determination of the specific mercury compounds present in the wastes is important for the understanding of mercury bioavailability in surrounding ecosystems. For example, the amount of various mercury compounds has an effect on mercury methylation and subsequent uptake by biota. Several mercury compounds are commonly formed during roasting of mercury ore including metacinnabar, corderoite ($Hg_3S_2Cl_2$), schuetteite ($Hg(SO_4)_2H_2O$), and mercury chlorides and oxides (Kim and others, 2000). All these compounds are more soluble than cinnabar, and as a result, they are more reactive and release Hg^{2+}, with the potential to form bioavailable methylmercury (Rytuba, 2000).

Mercury in acid-mine drainage and sediment

The mineralogy and geochemistry of the mercury deposits (table 2) are important factors in determining the pH and the composition of drainage downstream from the mines. For example, mine drainage from some silica-carbonate deposits is extremely acidic, as low as pH 2.2, owing to the presence of

Figure 12. Mercury deposits in California mercury mineral belt. Age in Ma (million years).

Table 2. Geologic and geochemical factors that control the composition of mine drainage from mercury deposits and mines in the California Coast Ranges.

	Silica-carbonate deposits	Hot-spring deposits
Trace metals	Ni-Co-Cr-Sb-Zn-Cu	Au-As-Sb-Li-W.
Alteration	Carbonates-quartz	Adularia-quartz-clays.
Sulfides	Pyrite and marcasite (5–50%)	Pyrite (2–5%).
Host rocks	Serpentinite, minor shale	Clastic rocks, lesser volcanic rocks.
Structural control	Serpentinite contacts	Faults and volcanic vents.
Ore minerals	Cinnabar, minor elemental Hg	Cinnabar.
Secondary minerals	Mercury sulfates and chlorides	Mercury sulfates and chlorides.

Figure 13. Typical mercury mine-waste calcines. Red-brown character of these mine wastes is a result of presence of iron oxide and fine-grained cinnabar.

large amounts of acid-water-generating iron-sulfide minerals. Acidic drainage from such mines is environmentally adverse because mercury and other metals are more mobile in low-pH conditions. Where acid-mine drainage flows through and reacts with mine wastes, soluble mercury compounds are leached, resulting in higher total mercury concentrations in water

(fig. 14). High concentrations of total mercury (as much as 450,000 ng/L (nanograms/liter)) and methylmercury (as much as 70 ng/L) are found in such mine drainage (see table 3, p. 34). The mercury and methylmercury concentrations in these mine waters are several orders of magnitude higher than uncontaminated baseline sites (table 3) and indicate that mercury mines in the California Coast Ranges are sites of significant mercury contamination.

In addition to acid-mine drainage, mercury mines with abundant iron sulfides also produce high concentrations of dissolved iron (> 8,000 mg/L (milligrams/liter)), which leads to precipitation of iron oxyhydroxides (Rytuba, 2000). Iron precipitates, as well as clay minerals, have a high capacity to sorb mercury; and as a result, they are an important source of mercury released from mercury-mine drainage. Such iron- and clay-rich sediment derived from mine wastes is the main source of mercury, where mercury contents are as high as 220 µg/g and methylmercury is as high as 110 ng/g. Particulate mercury is released from these sediments into streams during periods of high precipitation and resultant high runoff, which is common in the California winter climate. Particulate mercury released during high-flow becomes available to bacteria that methylate mercury later in the season, especially under the oxygen-depleted aquatic conditions typical in late summer. Methylmercury generated around these mine sites becomes bioavailable to organisms in the aquatic food web, especially fish. In some instances, mercury contents in fish collected downstream from some mercury mines exceed the 0.5 µg/g safe level for edible portions of fish established by the State of California (table 3). Mercury mines typically generate the highest concentrations of mercury-rich sediment and runoff with elevated mercury (some of which is acidic), and these sites are the primary sources of mercury that enters surrounding ecosystems in the California Coast Ranges.

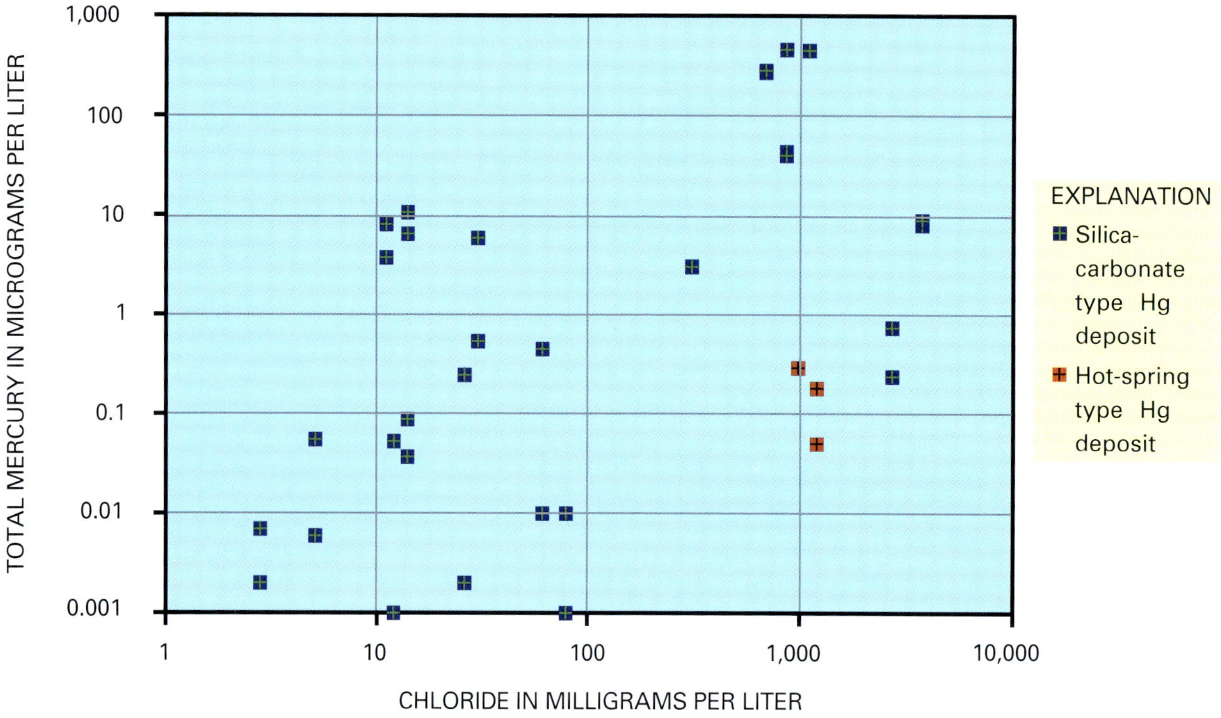

Figure 14. Total mercury concentration in acid-rock drainage from silica-carbonate and hot-spring-type mercury deposits in the California mercury mineral belt as a function of chloride concentration.

References Cited

Kim, C.S., Brown, G.E. Jr., and Rytuba, J.J., 2000, Characterization and speciation of mercury-bearing mine wastes using X-ray absorption spectroscopy (XAS): Science of the Total Environment, v. 261, p. 157–168.

Rytuba, J.J., 1996, Cenozoic metallogeny of California, in Coyner, A.R., and Fahey, P.L., eds., Geology and ore deposits of the American Cordillera: Geological Society of Nevada Symposium Proceedings, Reno/Sparks, Nev., April 1995, p. 803–822.

Rytuba, J.J., 2000, Mercury mine drainage and processes that control its environmental impact: Science of the Total Environment, v. 260, p. 57–71.

Rytuba, J.J., 2003, Mercury from mineral deposits and potential environmental impact: Environmental Geology, v. 43, p. 326–338.

The Southwestern Alaska Mercury Belt

By John E. Gray *and* Elizabeth A. Bailey

Abstract

Abandoned mercury mines are scattered over several thousand square kilometers in southwestern Alaska, primarily in the Kuskokwim River basin. Mercury ore is dominantly cinnabar, but elemental mercury is present at some mines. About 1,400 t of mercury have been produced from the region, but mines in the area have been closed since the 1970s. Stream-sediment samples collected downstream from the mines can contain total mercury concentrations as high as 5,500 µg/g. Such high mercury concentrations are related to the abundance of cinnabar, and in some instances minor elemental mercury, which are visible in streams below mine sites. Unfiltered mine-water samples contain total mercury as high as 2,500 ng/L; whereas, corresponding water samples filtered at 0.45 µm contain total mercury contents of less than 50 ng/L. These water data indicate that most of the mercury transported downstream from the mines is as finely suspended material rather than dissolved mercury. Although methylmercury contents (as much as 31 ng/g in sediments and 1.2 ng/L in stream water) represent only a small portion of the total mercury, these results indicate that part of the mercury is converted to bioavailable methylmercury. Muscle samples of fish collected downstream from mines contain total mercury concentrations as high as 0.62 µg/g (wet weight), of which 90 – 100 percent is methylmercury. However, the concentration of mercury in these fish is below the 1.0 µg/g action level for mercury in edible fish established by the U.S. Food and Drug Administration (FDA). Salmon contain total mercury contents of less than 0.1 µg/g and were the lowest mercury contents found for fish in the study, and well below the FDA action level.

Introduction

In addition to the large mercury mines in the California Coast Ranges, a much smaller belt of mercury mines and deposits is located in southwestern Alaska (fig. 15). Similar to some of the deposits in California, the mercury deposits in Alaska were formed near the Earth's surface in hot-spring environments (Gray and others, 1997). The Alaska mercury belt consists of numerous deposits and abandoned mines covering a wide area, mostly along the Kuskokwim River basin (fig. 15). Like most mercury mines worldwide, cinnabar is the dominant ore mineral, but native mercury is also found in a few localities. These mines are presently closed, but they produced about 41,000 flasks of mercury (1,400 t) from mining in the early 1900s through the 1970s. Although this mercury production is small compared with much larger mines throughout the world (fig. 5), the mines in southwestern Alaska produced more than 99 percent of all mercury mined in the State.

There are presently significant mine wastes containing cinnabar ore and minor amounts of elemental mercury near retorts

(fig. 16). Mercury remaining at these sites poses potential environmental hazards to the population and wildlife because mine drainage enters streams and rivers that are part of local ecosystems. To evaluate environmental concerns, the USGS measured concentrations of total mercury and methylmercury in stream sediment, soil, stream water, vegetation, and fish collected near these mines (Gray and others, 1996; Bailey and Gray, 1997; Gray and others, 2000; Bailey and others, 2002). Similar samples were also collected distant from the mines to compare total mercury and methylmercury concentrations in samples unaffected by mercury mining (table 3).

Stream-Sediment, Soil, and Vegetation Samples

Stream-sediment and soil samples collected near the mines in southwestern Alaska can contain total mercury concentrations as high as 5,500 µg/g (Gray and others, 1996; Bailey and Gray, 1997). Such high mercury concentrations are due to the abundance of cinnabar and minor amounts of elemental mercury present in these samples. Cinnabar is resistant to surface weathering and thus is common around these sites and in streams draining the mines. Concentrations of highly toxic methylmercury (as much as 41 ng/g in the stream-sediment and soil samples; table 3) are low relative to the high concentrations of total mercury in the stream-sediment and soil samples. Vegetation collected near the mines studied were also highly elevated in total mercury (as much as 970 ng/g) and methylmercury (as much as 11 ng/g) (Bailey and others, 2002). On a percentage basis, methylmercury generally composes less than one percent of the total mercury in the sediment, soil, and vegetation samples.

Stream Water and Fish

Stream waters draining the mercury mines are neutral to slightly alkaline, ranging in pH from 7.0 to 8.5. Acid-water-generating iron-sulfide minerals are rare, and as a result, near-neutral water pH is common around these mines. In addition, cinnabar is generally insoluble in water and does not readily form acid water during weathering. Thus, acidic mine water in streams is generally insignificant.

Unfiltered stream-water samples collected below the mines contained total mercury as high as 2,500 ng/L (fig. 17). Total mercury concentrations were several times higher in unfiltered stream water than in corresponding filtered-water samples, indicating that mercury transport is mostly as suspended particulates, probably particulate cinnabar. Most stream waters contained total mercury concentrations below the 2,000 ng/L drinking-water standard (fig. 17) recommended by the State of Alaska (Alaska Department of Environmental Conservation,

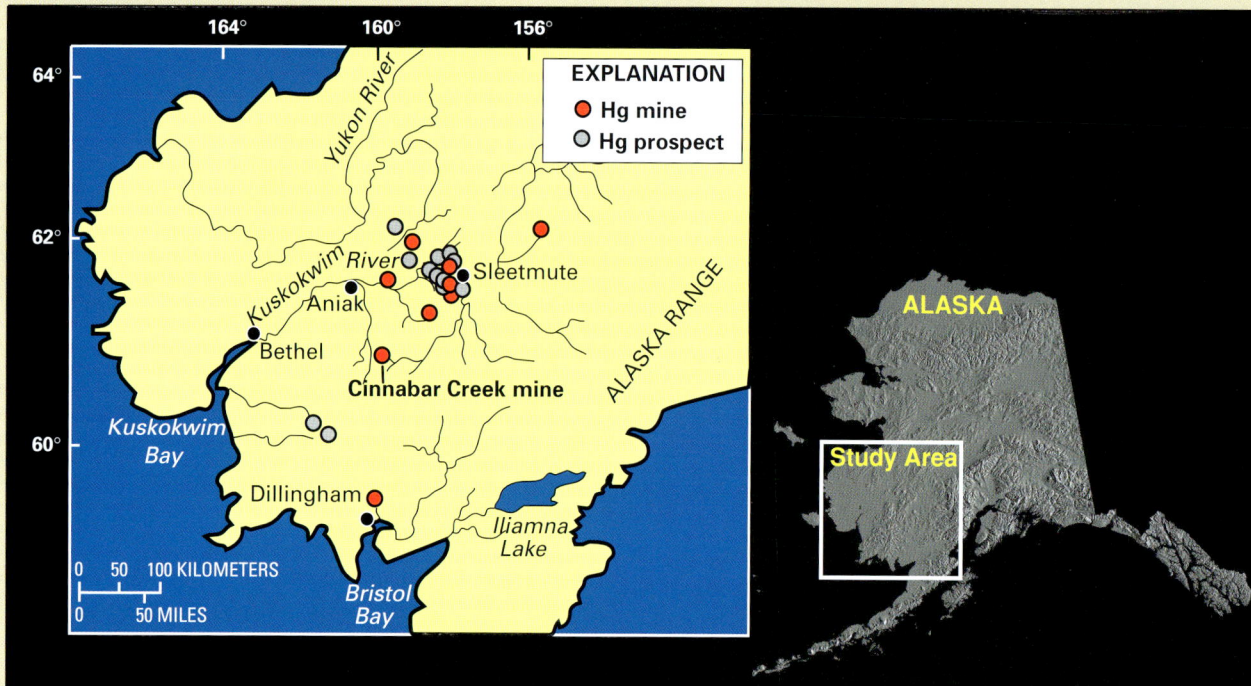

Figure 15. Location of mercury mines in southwestern Alaska.

Figure 16. Elemental mercury spilled at Red Devil mine retort, Alaska. Oxidation of elemental mercury to Hg^{2+} and subsequent methylmercury formation is a significant environmental concern around all mercury mines. The retort facility and nearby elemental mercury contamination have been removed from the Red Devil site and additional remediation efforts are ongoing.

1994), but exceed the 12 ng/L standard that the EPA has suggested may result in *chronic* effects to aquatic life (U.S. Environmental Protection Agency, 1992). As recommended by the EPA, when total mercury in stream water was found to exceed the 12 ng/L EPA chronic aquatic life standard, edible portions of fish were analyzed to determine their mercury contents (discussed in the next paragraph). Methylmercury concentrations in the stream-water samples were as much as 1.2 ng/L (table 3).

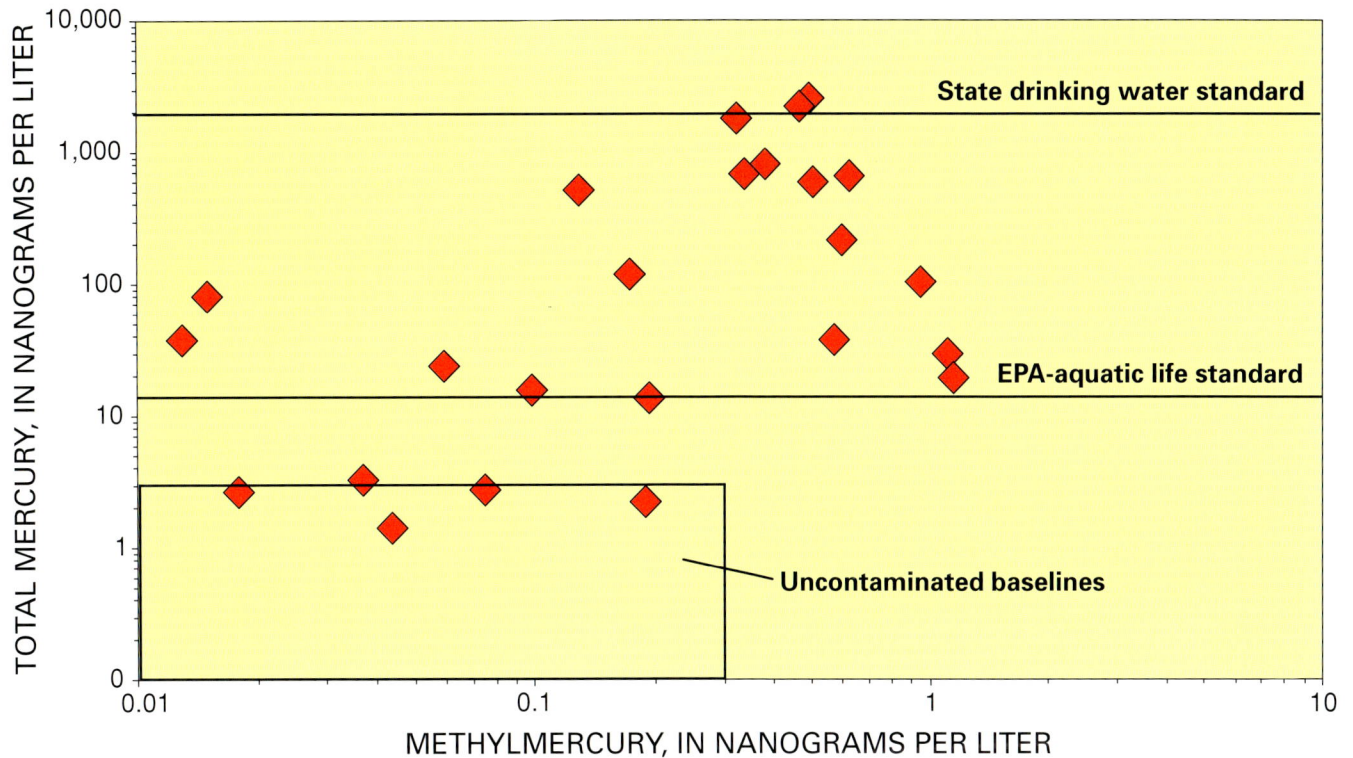

Figure 17. Concentration of mercury versus methylmercury in unfiltered water samples (red diamonds) collected from near mercury mines and uncontaminated baselines in southwestern Alaska. State of Alaska drinking water standard for mercury (2,000 ng/L) and EPA aquatic life mercury standard for adverse *chronic* effects to biota (12 ng/L) also shown for reference.

Similar to the results for the stream-sediment and soil samples, methylmercury concentrations in the stream-water samples constituted a small fraction, generally less than 3 percent of total mercury. However, methylmercury contents in unfiltered mine waters were generally higher than that found in unfiltered water from regional baseline sites uncontaminated by mercury mining (≤ 0.3 ng/L; table 3).

Samples of muscle from freshwater fish (fillets) collected near these mercury mines contained as much as 0.62 μg Hg/g (wet weight basis) (fig. 18). Of this, methylmercury makes up more than 90 percent of the total mercury (Gray and others, 2000), which is typical for most fish (National Academy of Sciences, 1978; U.S. Environmental Protection Agency, 1997). The mercury results for these fish indicate that part of the mercury is biologically available to the fish, especially fish collected nearest the mines. For example, fish collected near the Cinnabar Creek mine contained total mercury concentrations several times greater than mercury in fish collected distant from the mines. The State of Alaska has not established a regulatory standard for mercury contents in fish, and thus Alaska uses the Federal "action level" for mercury of 1.0 μg/g in edible portions of fish (fish muscle) established by the U.S. Food and Drug Administration (FDA) (Federal Register, 1979). All of the mercury contents in the fish collected in southwestern Alaska were below the FDA action level; when this concentration is exceeded, advisories are posted and the sale of fish is restricted. However, methylmercury contents in some fish collected from Cinnabar Creek (Gray and others, 2000) exceed the newly established standard

of 0.3 μg-methylmercury/g-fish (U.S. Environmental Protection Agency, 2001). Perhaps most importantly, all salmon collected by the USGS in this study contained total mercury concentrations less than 0.1 μg/g, the lowest mercury concentrations in this study, and below the recommended safe levels for mercury in fish (fig. 18). These results are significant because salmon are the most commonly consumed fish by residents and sport fishermen in the region.

Summary of the USGS Studies in Alaska

The concentration of total mercury is highly elevated especially in stream-sediment and soil samples collected from around the mercury mines in Alaska. These high mercury concentrations are related to the presence of cinnabar, which is a stable form of mercury with a low reactivity in water. Concentrations of methylmercury measured in the samples collected indicate that only minor conversion to this highly toxic form of organic mercury, but the elevated total mercury concentrations in fish collected near the mines indicate that some mercury is bioavailable to fish. Mercury concentrations in fish are useful for understanding the pathway of mercury in the food chain that can eventually affect humans. Although total mercury contents in sediment and water collected near the mines are elevated, all of the fish analyzed contained total mercury concentrations below the safe level for edible fish recommended by the FDA.

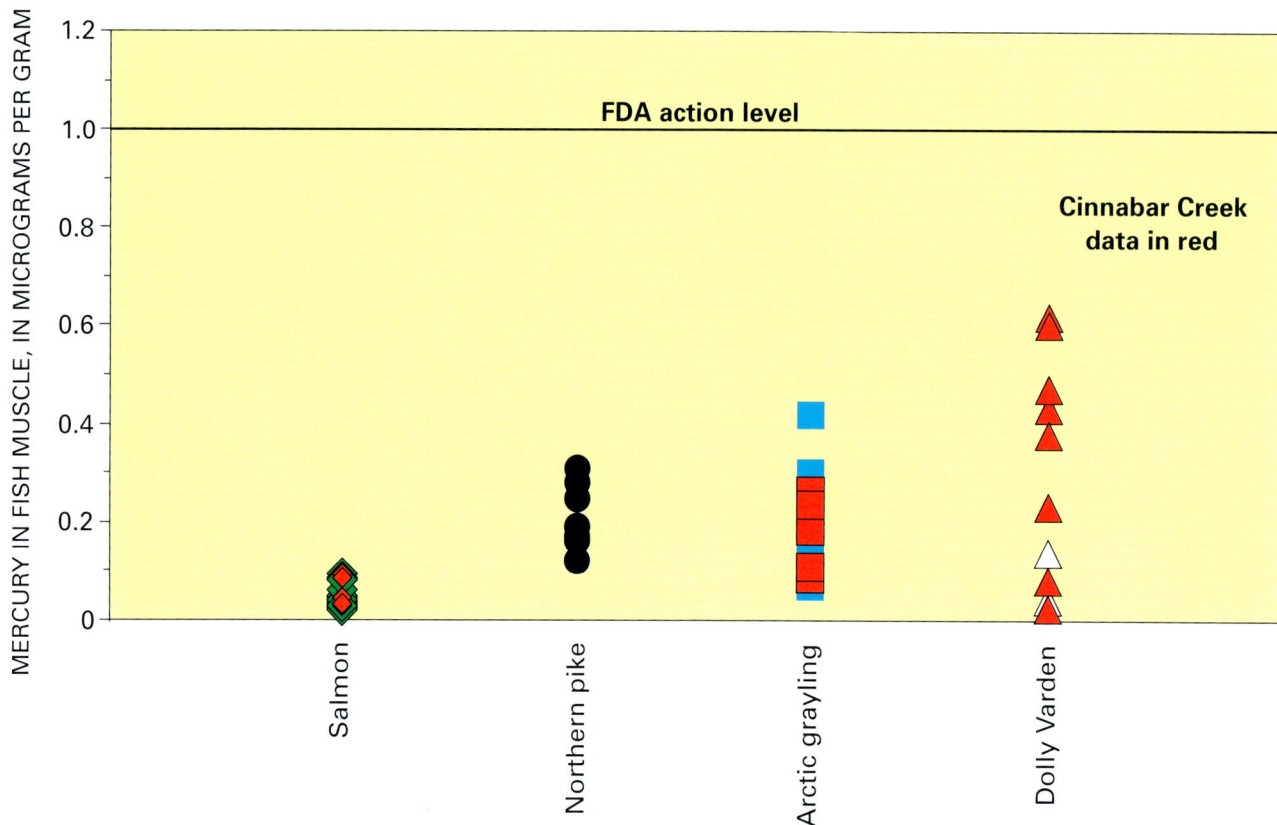

Figure 18. Mercury concentration in muscle for fish collected in southwestern Alaska. Fish collected nearest Cinnabar Creek mercury mine contain highest mercury concentrations. FDA action level is shown for reference.

References Cited

Alaska Department of Environmental Conservation, 1994, Drinking water regulations: State of Alaska, Department of Environmental Conservation Report 18-ACC-80, 195 p.

Bailey, E.A., and Gray, J.E., 1997, Mercury in the terrestrial environment, Kuskokwim Mountains region, southwestern Alaska, in Dumoulin, J.A., and Gray, J.E., eds., Geologic studies in Alaska by the U.S. Geological Survey, 1995: U.S. Geological Survey Professional Paper 1574, p. 41–56.

Bailey, E.A., Gray, J.E., and Theodorakos, P.M., 2002, Mercury in vegetation and soils at abandoned mercury mines in southwestern Alaska, USA: Geochemistry—Exploration, Environment, and Analysis, v. 2, p. 275–286.

Federal Register, 1979, Action level for mercury in fish, shellfish, crustaceans, and other aquatic animals: Comment from the Department of Health, Education, and Welfare, Food and Drug Administration, v. 44, no. 14, p. 3990–3993.

Gray, J.E., Gent, C.A., Snee, L.W., and Wilson, F.H., 1997, Epithermal mercury-antimony and gold-bearing vein deposits of southwestern Alaska: Economic Geology Monograph 9, p. 287–305.

Gray, J.E., Meier, A.L., O'Leary, R.M., Outwater, C., and Theodorakos, P.M., 1996, Environmental geochemistry of mercury deposits in southwestern Alaska—Mercury contents in fish, stream-sediment, and stream-water samples, in Moore, T.E., and Dumoulin, J.A., eds., Geologic studies in Alaska by the U.S. Geological Survey, 1994: U.S. Geological Survey Bulletin 2152, p. 17–29.

Gray, J.E., Theodorakos, P.M., Bailey, E.A., and Turner, R.R., 2000, Distribution, speciation, and transport of mercury in stream sediment, stream water, and fish collected near abandoned mercury mines in southwestern Alaska, U.S.A.: Science of the Total Environment, v. 260, p. 21–33.

National Academy of Sciences, 1978, An assessment of mercury in the environment: Washington, D.C., National Academy of Sciences, National Research Council, 185 p.

U.S. Environmental Protection Agency, 1992, Water quality standards; establishment of numeric criteria for priority toxic pollutants; states' compliance; final rule: Federal Register, 40 CFR Part 131, v. 57, no. 246, p. 60847–60916.

U.S. Environmental Protection Agency, 1997, Mercury study report to Congress, Volumes I–VIII: U.S. Environmental Protection Agency Report EPA-452/R-97-003.

U.S. Environmental Protection Agency, 2001, Water quality criterion for the protection of human health—Methylmercury: U.S. Environmental Protection Agency Report EPA-823-R-01-001.

Studies of Mercury Contamination from Gold Mining

Mercury Contamination from Hydraulic Gold Mining in the Sierra Nevada, California

By Michael P. Hunerlach *and* Charles N. Alpers

Abstract

Mercury contamination from the mining and recovery of gold during the late 19th and early 20th centuries is widespread in watersheds where there are historic placer-gold mines in the Sierra Nevada, California. Hydraulic mining has severely modified the region's geomorphology and hydrology, leading to increased turbidity of the natural waters, siltation of riverbeds, and contamination of the bottom sediments of reservoirs downstream of the mines. Both elemental mercury and methylmercury, which are potential risks to human health and to surrounding ecosystems, have been detected in the watersheds where hydraulic mining was widespread. Since 1998, the USGS has been characterizing specific "hot spots" in the Sierra Nevada to identify elevated concentrations of total mercury and methylmercury in water, soil, and biota. High levels of mercury bioaccumulation in various fauna, from mine sites and receiving waters, and visible elemental mercury in sediments and on bedrocks downstream of mines indicate a large part of the mercury used in gold ore processing was lost to the environment. The most elevated total mercury concentrations in the water and sediment were found in ground and tunnel sluices, the sites of historical gold recovery. Mercury bioaccumulation in fish in reservoirs and streams has prompted local officials to issue consumption advisories.

Introduction

Mercury contamination from historic placer-gold mines in the Sierra Nevada, California, represents a potential risk to human health and the environment (fig. 19). Elemental mercury that was used in the late 1800s and early 1900s for the recovery of gold at the mines and processing sites continues to enter local and downstream water bodies. Rivers, reservoirs, flood plains, and estuaries have been affected by the transport of mercury associated with contaminated sediments downstream from the mines. Since 1998, the USGS has been characterizing specific "hot spots" in the Sierra Nevada to identify elevated concentrations of total mercury and methylmercury in water, soil, and biota. Unfiltered mine waters sampled in 1998 contained total mercury concentrations ranging from 40 to 10,400 ng/L, and concentrations of methylmercury in corresponding unfiltered water samples ranged from 0.01 to 1.1 ng/L. In addition, samples of sluice-box sediments contained total mercury concentrations ranging from 600 to 26,000 µg/g.

Although these sluice-box sediments are highly elevated in mercury, sediments collected from the Sacramento River farther downstream from the mines show significant dilution of mercury with concentrations similar to those found in sediments collected from uncontaminated baseline sites in North America (table 3). Based on these studies, the USGS estimates that hundreds to thousands of pounds of elemental mercury may remain at each of numerous sites affected by hydraulic placer-gold mining in the Sierra Nevada (Hunerlach and others, 1999). Total mercury concentrations in muscle samples of black bass (*Micropterus* spp.), including largemouth, smallmouth, and spotted bass collected from areas affected by historic gold mining in the Sierra Nevada, ranged from 0.20 to 1.5 µg/g (wet weight basis) in five Sierra Nevada reservoirs affected by historic hydraulic gold mining (table 3); (May and others, 2000). The mercury content in many of these fish exceeded the 1.0 µg/g FDA action level and the 0.5 µg/g safe level used by the State of California (table 3). Based on USGS studies, a better understanding is emerging of mercury distribution, ongoing transport, transformation processes, and the extent of biological uptake in areas affected by gold mining in the Sierra Nevada (Hunerlach and others, 1999; May and others, 2000).

Origins of Hydraulic Mining in California

Hydraulic mining began in California between 1852 and 1853, shortly after the discovery of gold. Vast gravel deposits in rivers within the Sierra Nevada gold belt contained large quantities of placer gold that provided the basis for the first large-scale mining in California. California had all the essential materials for the cheap and efficient method of hydraulic mining. Water was abundant, vast Tertiary-age gravels were rich with fine-grained gold, and elemental mercury (used for gold recovery) was being produced extensively in the Coast Ranges mercury mines (Alpers and Hunerlach, 2000; Rytuba, this volume). Hydraulic mining used high-pressure water spraying (fig. 20) to deliver large volumes of water that stripped the ground of soil, sand, and gravel above bedrock. The water and sediment formed slurries that were directed through large sluice boxes at sites near the discovery of gold (fig. 21), where the gold was recovered. An extensive water transfer system of ditches, canals, and vertical pipes was constructed to provide the sustained water pressure necessary for hydraulic mining. As mining progressed into deeper gravels, tunnels were constructed to facilitate drainage and to provide an exit route for mining debris from the bottom of hydraulic mine pits. The

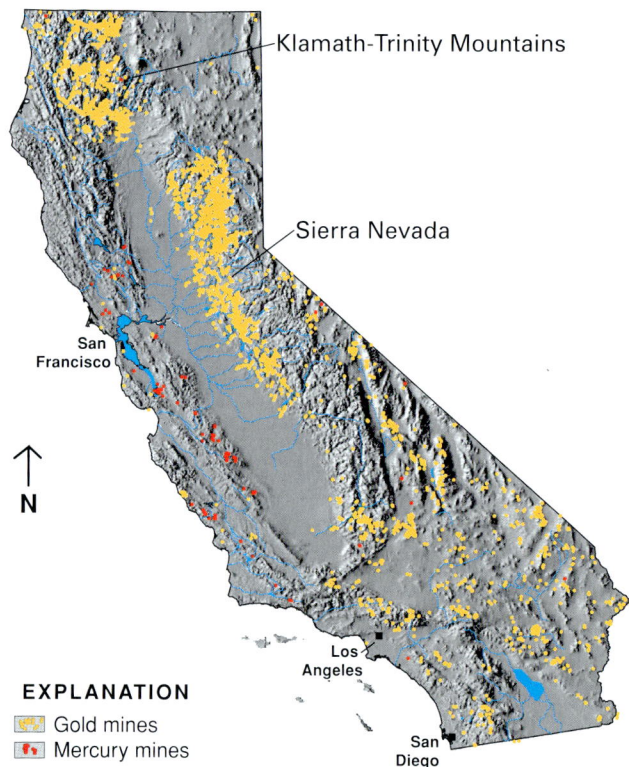

Figure 19.   Location of gold and mercury mines in California.

Elemental Mercury Use in Hydraulic Mining

The capability of elemental mercury to alloy or amalgamate with gold has been well known for more than 2,000 years. Miners used elemental mercury to recover gold throughout the western United States at both placer (alluvial) and hardrock (lode) mines. The vast majority of elemental mercury lost to the environment in California was from placer-gold mines, which used hydraulic, drift, and dredging methods to process more than 5.5 billion cubic yards of gold-bearing gravels. In placer mine operations, loss of elemental mercury during gold recovery was reported to be as much as 30 percent or higher, depending upon the efficiency of the gold recovery apparatus (Averill, 1946). More than 100,000 t of mercury was produced in California since 1850, of which more than 10,000 t was used to extract gold by amalgamation from the gold-bearing gravels (Churchill, 1999).

In a typical sluice system, hundreds of pounds of elemental mercury were added to riffles and troughs to enhance gold recovery. The density of elemental mercury is between that of gold and the gravel slurry, so gold and gold-quicksilver amalgam would sink, while the sand and gravel would pass over the elemental mercury and through the sluice. Gravel and cobbles that entered the sluices caused the elemental mercury to flour, or break into small particles. Flouring was aggravated by agitation, exposure of elemental mercury to air, and other chemical reactions. Eventually, the entire bottom of the sluice became coated with elemental mercury. Some mercury escaped from the sluice through leakage and was transported downstream with the placer tailings. Because such large volumes of turbulent water flowed through the sluice, many of the finer grained gold particles attached to elemental mercury particles were washed through and out of the sluice before they could settle in the riffles laden with elemental mercury. A modification of the sluicing technique known as an undercurrent (fig. 23) was developed to address this loss. Fine-grained sediment was allowed to drop onto the undercurrent, where gold and gold-mercury amalgam were caught.

As a result of the extensive use of mercury for amalgamation during gold recovery and its subsequent loss, elemental mercury is commonly present in riverbanks, soils, and drainages throughout the region of historic gold mining operations. Mercury concentrations in sediments are generally higher in areas of large-scale gold mining and processing activities. In sluice boxes, where gold was recovered, and in areas where mining debris is continually reworked by seasonal runoff, total mercury concentrations can be as much as 1,000 µg/g in tailings. Farther downstream, the San Francisco Bay is the recipient of more than 150 years of contaminated sediment transport, where close correlation exists between total mercury concentrations and percentage of fine-grained sediments in the bay (Hornberger and others, 1999). In general, total mercury concentrations tend to increase with the amount of fine-grained material because the amount of surface area available for adsorption increases with an increase in the amount of fine-grained material. Throughout the Sierra Nevada millions of cubic yards of both coarse- and fine-grained placer tailings are subject to continued mercury remobilization from either natural or anthropogenic effects.

tunnels provided a protected environment for sluices and a way to discharge processed sediments (placer tailings) to adjacent waterways (fig. 22).

Hydraulic mines operated on a large scale from the 1850s to the 1880s in California's northern Sierra Nevada region displacing a total of more than 1.6 billion cubic yards of sediment. In 1884, an important legal judgment (the Sawyer Decision) prohibited discharge of mining debris in the Sierra Nevada region (Gilbert, 1917), but not in the Klamath-Trinity Mountains (fig. 19), where hydraulic mining continued until the 1950s. Hydraulic mining spread quickly throughout the western United States gold mining districts and continues today on a limited permit basis n Alaska, although elemental mercury is rarely used for gold recovery in the United States.

Underground mining of placer deposits (drift mining) and of hardrock gold-quartz vein deposits produced most of California's gold from the mid-1880s to the early 1900s. Dredging of gold-bearing sediments in the Sierra Nevada foothills has been an important source of gold since the early 1900s. Elemental mercury was used extensively until the early 1960s in the dredging of large flood-plain deposits of gravel and topsoil (Alpers and Hunerlach, 2000). Elemental mercury lost during historic gold mining is recovered today as a byproduct from large- and small-scale dredging operations in many placer districts throughout the United States.

Figure 20.   Water cannons used tremendous volumes of water under high pressure to break down the gold-bearing gravel deposits in the Sierra Nevada (Malakoff Diggings, about 1860). (Photograph courtesy of California Department of Parks and Recreation.)

Figure 21.   Gravel deposits were washed into sluices where gold was recovered by gravity separation. Amalgamation with elemental mercury was then used to extract the gold (about 1850). (Photograph courtesy of Siskiyou County Historical Society.)

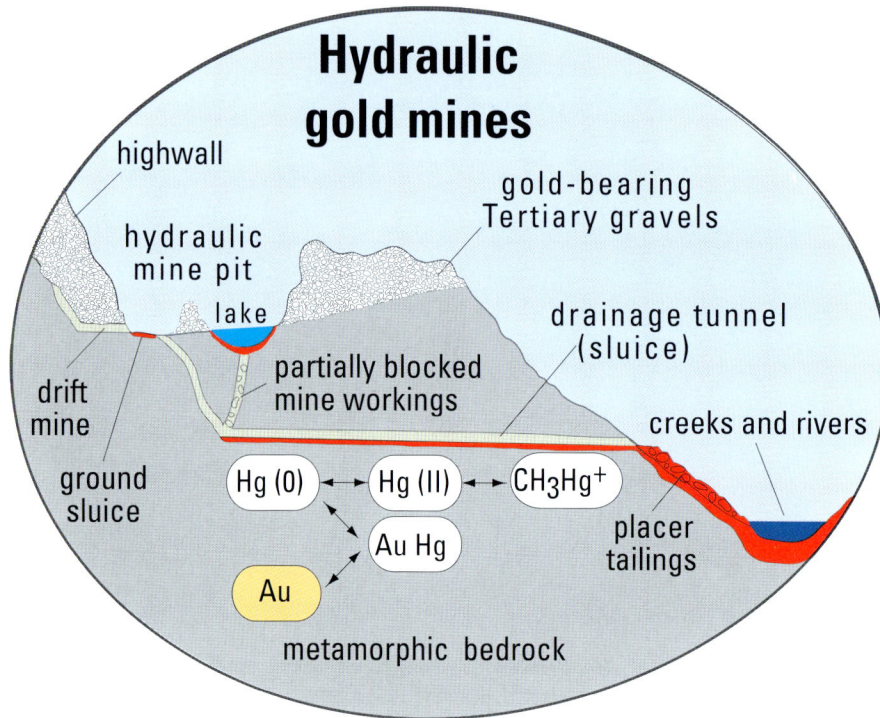

Figure 22. Schematic diagram showing transport of mercury and placer tailings from a hydraulic mine pit through a drainage tunnel and discharge into creeks and rivers.

Figure 23.   View of sluice system, Siskiyou County, California (about 1860). (Photograph courtesy of Siskiyou County Historical Society.)

References Cited

Alpers, C.N., and Hunerlach, M.P., 2000, Mercury contamination from historic gold mining in California: U.S. Geological Survey Fact Sheet FS-061-00, 6 p.

Averill, C.V., 1946, Placer mining for gold in California: California State Division of Mines and Geology Bulletin 135, 336 p.

Churchill, R., 1999, Insights into California mercury production and mercury availability for the gold mining industry for the historical record: Geological Society of America Abstracts with Programs, v. 31, no. 6, p. 45.

Gilbert, G.K., 1917, Hydraulic-mining debris in the Sierra Nevada: U.S. Geological Survey Professional Paper 105, 155 p.

Hornberger, M.I., Luoma, S.N., van Geen, A., Fuller, C., and Anima, R., 1999, Historical trends of metals in the sediments of San Francisco Bay, California: Marine Chemistry, v. 64, p. 39–55.

Hunerlach, M.P., Rytuba, J.J., and Alpers, C.N., 1999, Mercury contamination from hydraulic placer-gold mining in the Dutch Flat mining district, California: U.S. Geological Survey Water-Resources Investigations Report 99-4018B, p. 179–189.

May, J.T., Hothem, R.L., Alpers, C.N., and Law, M.A., 2000, Mercury bioaccumulation in fish in a region affected by historic gold mining— The South Yuba River, Deer Creek, and Bear River watersheds, California, 1999: U.S. Geological Survey Open-File Report 00-367, 30 p.

Mercury in the Carson River Basin, Nevada

By Stephen J. Lawrence

Abstract

The Carson River from Carson City, Nevada, to the Carson Sink is one of the most severe cases of mining-related mercury contamination in the United States. Elemental mercury was used to extract gold and silver in ore mined between 1863 and 1900 from the Comstock Lode near Virginia City, Nevada. During this time, about 7,000 t of elemental mercury was lost to the environment in spent mine tailings contaminated with mercury. These tailings and associated elemental mercury were eroded, transported, and dispersed throughout the lower Carson River, Lahontan Reservoir, and the Carson Sink by floods that occurred 19 times between 1861 and 1997. Total mercury concentrations in Lahontan Reservoir bottom sediments were as much as 80 µg/g and 100 µg/g in deep-water and deltaic sediments, respectively. Total mercury concentrations in unfiltered water samples from the Carson River were as much as 28 µg/L. Methylmercury concentrations in bottom sediments of the Carson River and Lahontan Reservoir were as much as 29 ng/g, whereas methylmercury contents in unfiltered water were as much as 21 ng/L. Fish collected from the lower Carson River and Lahontan Reservoir contained as much as 16 µg/g of total mercury in their tissues, and crayfish contained as much as 50 µg Hg/g.

Introduction

The Carson River in Nevada presents one of the most severe cases of mining-related mercury contamination in the United States. As much as 7,000 t of elemental mercury, which was used to extract gold and silver ores, is estimated to have been lost to the Carson River basin during the Comstock Lode mining period beginning in the 1850s (Smith, 1943). Elemental mercury lost or discarded during mining of the Comstock Lode has contaminated sediments of the Carson River (figs. 24 and 25).

Little thought was given to the potential environmental effects of mercury in the Carson River until the 1970s. In 1973, the USGS completed the first assessment of mercury in the Carson River (Van Denburgh, 1973). Since Van Denburgh (1973), additional studies have evaluated the association of mercury with sediment and organic material, its movement within the river, mercury methylation, and mercury accumulation in aquatic organisms in this ecosystem (for example, Gustin and others, 1994; Miller and others, 1995; Bonzongo and others, 1996; Wayne and others, 1996; Lechler and others, 1997; Hoffman and Taylor, 1998; Marvin-DiPasquale and Oremland, 1999). As a result of these studies, the Carson River, including the Stillwater Wildlife Refuge, Stillwater Wildlife Management Area, and Fallon Wildlife Refuge, was listed on the EPA National Priorities List in 1990 as the Carson River Mercury Superfund site (fig. 24).

History of the Comstock Lode Mining Period

The Comstock Lode was discovered in the spring of 1859 (Smith, 1943). Two groups of prospectors, working within a mile of each other, found gold-bearing rock near the areas that eventually became the towns of Virginia City and Gold Hill. These discoveries became the richest silver lode in the United States. The lode took its name from Henry Comstock, who gained a share of the most famous discovery, called the Ophir mine. Initially, placer mining was used to extract gold from gravel deposits despite a troublesome blue mud that was later identified as silver. Eventually, placer deposits played out and miners began using hardrock methods such as underground tunneling in order to follow rich gold ore. By 1863, the discoveries by individual prospectors had become the property of shrewd businessmen, such as James Fair and John Mackay, who had business, political, and mining knowledge. Mining companies and corporations were formed and generated operating capital by selling mining stocks to investors as far away as San Francisco and New York.

Soon after hardrock mining began, stamp mills were rapidly constructed to process the ore; and by 1863, 66 stamp mills were operating in the Carson River basin (fig. 26), primarily from Carson City to about 6 km downstream from Dayton (fig. 24). These mills used a mechanized amalgamation system, called the Washoe process, to rapidly extract gold and silver from the ores. In this process, rock is finely ground, then mixed with elemental mercury, and steam heated in pans to eliminate the sulfides in the ore that inhibit the recovery of gold and silver. The heat also vaporized mercury from the gold and silver amalgam, leaving gold and silver concentrates behind. The vaporized mercury was collected, cooled and condensed back to the liquid form, and collected for reuse. The remaining rock slurry, which contained small amounts of elemental mercury, was discarded either to the river or to tailings ponds near the mills. For every ton of ore processed using the Washoe method, as much as 1.5 pounds of elemental mercury was lost in the tailings (Smith, 1943). In the late 1800s, cyanide leaching began to replace mercury amalgamation as the preferred extraction method because of the higher rate of success of gold and silver recovery. Beginning about 1901, cyanide leaching became widely used in the basin.

More than 16,000,000 t of ore were estimated to have been removed from mines of the Comstock Lode, and about 70,000,000 oz of silver (2,500 kg) and 5,000,000 oz of gold (180 kg) were produced (Smith, 1943). The monetary value of gold

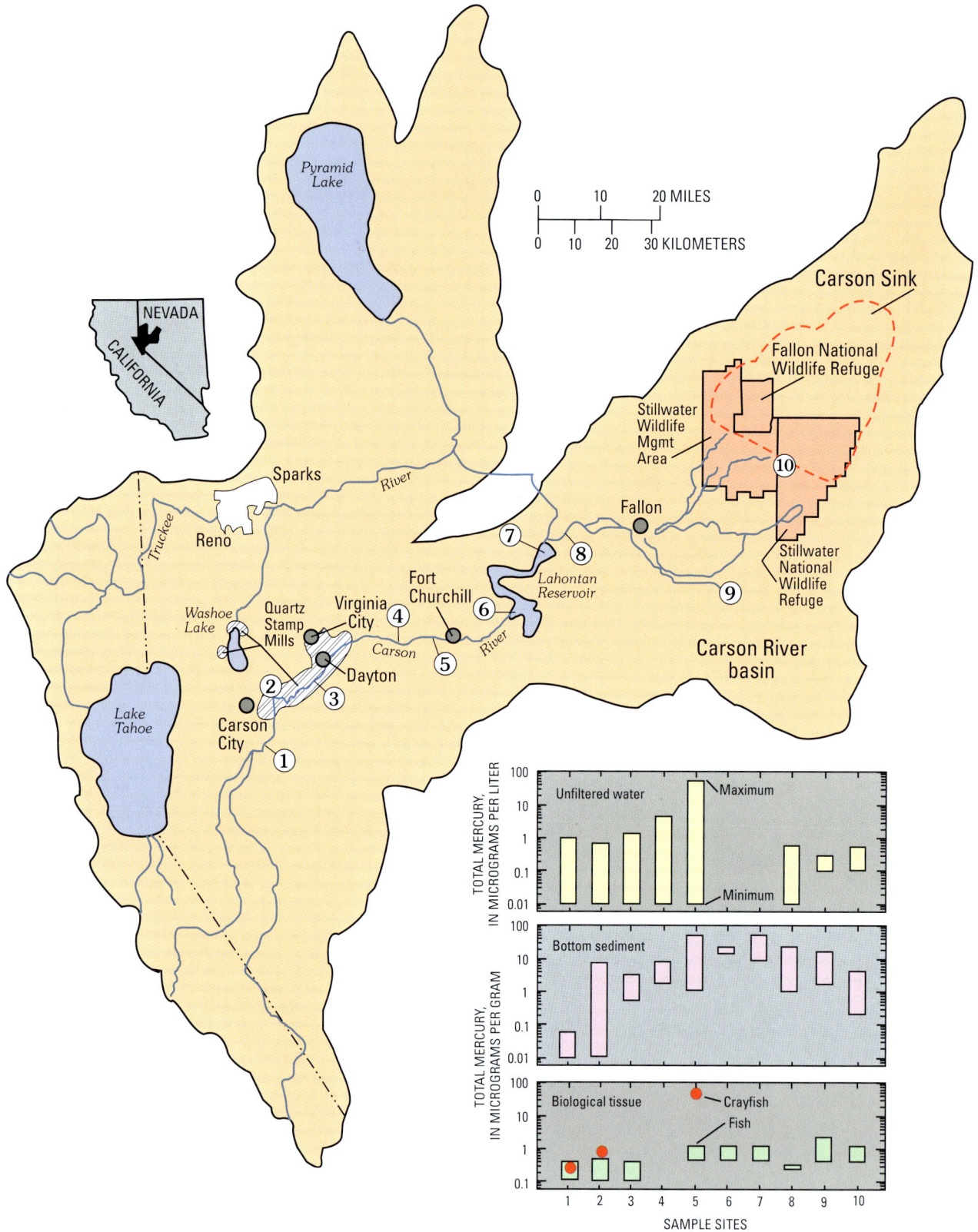

Figure 24. Location of sample sites in the Carson River basin, Nevada, and total mercury concentrations in unfiltered water, bottom sediment, and biological tissues collected from the Carson River.

Figure 25.   View of Carson River in Dayton Valley following flooding in January 1997. (Photograph by Pat Glancy, U.S. Geological Survey.)

and silver recovered (1859–1920) was about $350,000,000 (1920 dollars)(Smith, 1943).

Mercury in Soils, Bottom Sediment, Water, and Fish

The Carson River flooded 19 times between 1861 and 1997. These floods eroded, transported, and dispersed mercury-bearing mine tailings throughout the Carson River basin. Mine tailings and mercury-bearing stream-bottom sediments present throughout the basin contain total mercury concentrations as high as 1,610 µg/g (table 3). Total mercury concentrations exceeding 25 µg/g are common in flood-plain soils near Dayton, Fort Churchill, and the Carson Sink, and greater than 500 µg/g on the alluvial fans where Gold and Six Mile Creeks meet the Carson River (Hoffman and others, 1989).

Because of alluvial dispersion and dilution, the concentration of total mercury in sediment and soil becomes progressively lower farther downstream from contaminated tailings. Where soil washes into the river and mixes with bottom sediment, the total mercury concentrations measured in these sediments decline by nearly an order of magnitude. In 1998, total mercury concentrations in river-bottom sediments were significantly higher at sites near Fort Churchill (60 µg/g, site 5; fig. 24) compared with sites upstream from Carson City (0.01 µg/g, site 1; fig. 24), where the farthest upstream stamp mills were located. At Fort Churchill, the river is actively eroding sediments that were deposited on flood plains in the last century. Before 1915, mercury-contaminated sediments were deposited in the Carson Sink, a large, natural evaporation basin for Carson River water. After 1915, when the Lahontan Dam was completed, all mercury-contaminated sediment collected in the Lahontan Reservoir. Deep-water-bottom and deltaic-bottom sediments contain most of the mercury in the reservoir. The concentration of total mercury is as much as 80 µg/g in deep-water-bottom sediments and as much as 100 µg/g in deltaic-bottom sediments in Lahontan Reservoir. USGS research suggests that Lahontan Reservoir acts as an imperfect sediment trap. Mercury-laden sediment escapes the reservoir, especially during large episodic floods such as the flood of January 1997 (fig. 25); (Hoffman and Taylor, 1998).

Figure 26. Vivian quartz mill on Carson River (about 1870). (Photograph courtesy of Nevada Historical Society.)

Mercury is present primarily in its elemental form in tailings and bottom sediments between Carson City and Dayton (Lechler and others, 1997). USGS research suggests that as these tailings and sediment move downstream toward Lahontan Reservoir, the elemental mercury is absorbed onto clays, organic matter, and iron and manganese coatings. In the lower Carson River and in Lahontan Reservoir, methylmercury concentrations in sediments are as much as 29 ng/g, significantly higher than that in uncontaminated baseline sediments (table 3). These high methylmercury concentrations are probably related to the oxidation of elemental mercury and high organic-carbon contents, which are favorable for mercury methylation (Hoffman and others, 1989). Unfiltered water samples generally contain total mercury at concentrations similar to those in bottom sediments, particularly when flow in the Carson River exceeds 1,000 cubic feet per second (cfs). At this and higher stream flows, the river transports sediment particles (often mercury-laden) both from the stream bottom and from eroding banks (Hoffman and Taylor, 1998). Total mercury concentrations in unfiltered water samples collected from sites near Carson City are similar to those in samples from near Dayton (fig. 24). However, unfiltered water samples collected from Fort Churchill contain total mercury concentrations as high as 28 µg/L (table 3). Methylmercury

contents in unfiltered water collected from Lahontan Reservoir were as much as 7.8 ng/L (Hoffman and Thomas, 2000) and as much as 21 ng/L (Gustin and others, 1994). Similar to the results for bottom sediments, mercury and methylmercury contents in unfiltered water samples collected from the Carson River system were significantly higher than that from uncontaminated baseline sites throughout the United States (table 3).

Total mercury concentrations in the tissues of crayfish and various fish species show downstream increases that parallel those in water and bottom-sediment mercury concentrations (fig. 24). Total mercury concentrations in fish tissue (walleye) in Lahontan Reservoir were found to be as high as 16 µg/g (U.S. Environmental Protection Agency, 2002; Wayne Praskins, U.S. Environmental Protection Agency, Region 9, oral commun., December 2002), which greatly exceeds the FDA action level for Hg (1.0 µg/g) for human consumption of fish (table 3). A total mercury concentration of about 50 µg/g was measured in whole crayfish during a severe drought in 1992 at Fort Churchill (fig. 24). The Nevada Division of Wildlife has issued an advisory against consumption of fish from the lower Carson River and Lahontan Reservoir. Due to the severity of mercury contamination, the USGS continues to monitor and study the Carson River.

References Cited

Bonzongo, J.C., Heim, K.J., Warwick, J.J., and Lyons, W.B., 1996, Mercury levels in surface waters of the Carson River-Lahontan Reservoir system, Nevada—Influence of historic mining activities: Environmental Pollution, v. 92, p. 193–201.

Gustin, M.S., Taylor, G.E., Jr., and Leonard, T.L., 1994, High levels of mercury contamination in multiple media of the Carson River drainage basin of Nevada—Implications for risk assessment: Environmental Health Perspectives, v. 102, no. 9, p. 772–778.

Hoffman, R.J., Hallock, R.J., Rowe, T.G., Lico, M.S., Burge, H.L., and Thompson, S.P., 1989, Reconnaissance investigation of water quality, bottom sediment, and biota associated with irrigation drainage in and near Stillwater Wildlife Management area, Churchill County, Nevada, 1986 – 87: U.S. Geological Survey Water Resources Investigations Report 89-4105, 150 p.

Hoffman, R.J., and Taylor, R.L., 1998, Mercury and suspended sediment, Carson River basin, Nevada—Loads to and from Lahontan Reservoir in flood year 1997 and deposition in reservoir prior to 1983: U.S. Geological Survey Fact Sheet FS-001-98, 6 p.

Hoffman, R.J., and Thomas, K.A., 2000, Methylmercury in water and bottom sediment along the Carson River system, Nevada and California, September 1998: U.S. Geological Survey Water Resources Investigations Report 00-4013, 17 p.

Lechler, P.J., Miller, J.R., Hsu, L.C., and Desilets, M.O., 1997, Mercury mobility at the Carson River Superfund Site, west-central Nevada, USA—Interpretation of mercury speciation data in mill tailings, soils, and sediments: Journal of Geochemical Exploration, v. 58, p. 259–267.

Marvin-DiPasquale, M., and Oremland, R.S., 1999, Microbial mercury cycling in sediments of the Carson River system, *in* Sakrison, R., and Sturtevant, S., eds., Watershed management to protect declining species: Proceedings of the Water Resources Conference, Seattle, Wash., Dec. 5–9, 1999, p. 517–520.

Miller, J.R., Lechler, P.J., Rowland, J., Desilets, M., and Hsu, L.C., 1995, An integrated approach to the determination of the quantity, distribution, and dispersal of mercury in Lahontan Reservoir, Nevada, U.S.A.: Journal of Geochemical Exploration, v. 52, p. 45–55.

Smith, G.H., 1943, The history of the Comstock Lode, 1850–1920: Nevada University Bulletin, v. 37, no. 3, 305 p.

U.S. Environmental Protection Agency, 2002, National priorities list sites in Nevada, Carson River mercury site: U.S. Environmental Protection Agency, www.epa.gov/superfund/sites/npl/nv.htm, accessed 12/17/02.

Van Denburgh, A.S., 1973, Mercury in the Carson and Truckee River basins of Nevada: U.S. Geological Survey Open-File Report 73-352, 14 p.

Wayne, D.M., Warwick, J.J., Lechler, P.J., Gill, G.A., and Lyons, W.B., 1996, Mercury contamination in the Carson River, Nevada—A preliminary study of the impact of mining wastes: Water, Air, and Soil Pollution, v. 92, p. 391–408.

Volcanic Emissions of Mercury

By Todd K. Hinkley

Abstract

Measurement of mercury emissions from some representative volcanoes has aided in understanding volcanic contributions to the overall atmospheric budget of mercury. Present estimates suggest that all volcanoes worldwide contribute about 60 t of mercury to the atmosphere each year. Volcanic sources of mercury to the atmosphere are small on a global scale.

Introduction

Emissions from volcanoes have been known for many years as a source of mercury to the atmosphere. Chemical constituents, including mercury, emanating from quiescently degassing (or non-explosive) volcanoes (fig. 27) have been measured by scientists in recent decades. The amount of mercury contributed to the atmosphere from this natural geological source may assist in understanding (1) the relative contributions and total amount of mercury in the atmosphere that comes from volcanoes and other natural sources; and (2) how much of the total mercury in the environment in pre-industrial times came from volcanoes,

when the total amount of mercury in the air may have been lower.

Mercury, as well as many other trace elements and gases, is emitted and put into the air by volcanoes. In addition to mercury, surprisingly large amounts of several toxic elements, including lead, cadmium, and bismuth, are present in the plumes of volcanoes. The USGS has been involved in efforts to determine the amount of some of those metals emitted by volcanoes on a worldwide basis (Hinkley and others, 1999).

Measuring Mercury Emissions from Volcanoes

Efforts to measure the amount of mercury emitted from quiescently degassing volcanoes have centered on Mount St. Helens (Washington, U.S.A.), White Island (New Zealand; fig. 28), and Kilauea (Hawaii, U.S.A.). The collection of mercury emitted from volcanoes requires specialized sampling apparatus (fig. 29). Methods for sample preparation and measurement have been recently refined for better quantitative measurement of total mercury (Vandal and others, 1993; Ferrara and others,

Figure 27. Vapor and steam emanating from Volcano Farallon de Pajaros, Mariana Islands (plane wing in foreground). USGS scientists are currently measuring mercury emissions from several volcanoes worldwide.

1994). Because of the complexity of sampling, costs, and safety issues, mercury emissions have been measured only during limited time intervals at a few quiescent volcanoes; continuous monitoring for mercury emissions at all volcanoes worldwide is not possible. However, the total output of sulfur from volcanoes worldwide has been reliably estimated. If both mercury and sulfur emissions are measured during sampling at several volcanoes (Ferrara and others, 1994), and these measurements are assumed to be representative of that from volcanoes throughout the world, it is possible to obtain a reliable estimate for worldwide volcanic emissions of mercury. Based on these measurements, mercury emissions from volcanoes are approximately 1/1,000,000 of the amount of sulfur emitted, although at a few volcanoes the fraction of mercury is larger, perhaps 1/10,000.

Explosive volcanic eruptions also inject mercury into the atmosphere, adding significantly to the total mass of mercury put into the air by volcanic sources. However, because of the difficulty of taking such measurements safely, the amount of mercury coming from explosive eruptions is more difficult to estimate accurately than that from quiescently degassing volcanoes. Explosive volcanoes may contribute as much mercury as is emitted by quiescent volcanoes (Varekamp and Buseck, 1986).

Worldwide Contribution of Mercury from Volcanoes

The total contribution of mercury to the atmosphere from all sources worldwide is estimated to be about 6,000 t/year (Fitzgerald, 1986). Mercury emissions from quiescent volcanoes are estimated to be about 25–30 t/year worldwide (Varekamp and Buseck, 1986; Fitzgerald, 1986). Thus, mercury emissions from volcanoes probably account for less than one percent of the total global contribution of mercury to the atmosphere. If the amount of mercury emitted from explosive volcanic eruptions is also considered, the fraction is larger, but the total volcanic output of mercury is probably less than 60 t/year (table 1). In pre-industrial times, mercury emitted from volcanoes was probably similar to the amount today, but because the total amount of mercury contributed to the atmosphere was smaller in pre-industrial times, the volcanic contribution was a larger portion of the total. In fact, the amount of other trace metals emitted by volcanoes in pre-industrial times has been shown by the USGS to account for most of the total mass of a suite of volatile trace metals that were deposited in annual layers of ice preserved in the Antarctic ice sheet (Matsumoto and Hinkley, 2001).

Figure 28.   Volcanic emissions from White Island, New Zealand.

Figure 29. USGS and university scientists collecting volcanic gas samples for the measurement of mercury and other chemical constituents.

References Cited

Ferrara, R., Maserti, B.E., De Liso, A., Cioni, R., Raco, B., Taddeucci, G., Edner, H., Ragnarson, P., Svanberg, S., and Wallinder, E., 1994, Atmospheric mercury emission at Solfatara Volcano (Pozzuoli, Phlegraean fields, Italy): Chemosphere, v. 29, no. 7, p. 1421–1428.

Fitzgerald, W.F., 1986, Cycling of mercury between the atmosphere and oceans, in Buat-Menard, P., ed., The role of air-sea exchange in geochemical cycling: Dordrecht, Reidel Publishing Co., NATO, Advanced Science Institutes Series, p. 363–408.

Hinkley, T.K., Lamothe, P.J., Wilson, S.A., Finnegan, D.L., and Gerlach, T.M., 1999, Metal emissions from Kilauea, and a suggested revision of the estimated worldwide metal output by quiescent degassing of volcanoes: Earth and Planetary Science Letters, v. 170, no. 3, p. 315–325.

Matsumoto, A., and Hinkley, T.K., 2001, Trace metal suites in 75,000 years of Antarctic ice are consistent with emissions from quiescent degassing of volcanoes worldwide: Earth and Planetary Science Letters, v. 186, p. 33–43.

Vandal, G.M., Fitzgerald, W.F., Boutron, C.F., and Candelone, J.P., 1993, Variations in mercury deposition to the Antarctic over the past 34,000 years: Nature, v. 362, p. 621–623.

Varekamp, J.C., and Buseck, P.R., 1986, Global mercury flux from volcanic and geothermal sources: Applied Geochemistry, v. 1, p. 65–73.

Summary

By John E. Gray

The studies presented in this volume outline a few sources of potential contamination of mercury to air, water, land, and biota. In the United States and throughout the world, emissions from coal-fired electrification power plants contribute a significant proportion of the mercury found in the atmosphere (table 1). Numerous research studies are attempting to evaluate whether mercury in these power plant emissions contributes to environmental contamination of the air, water, and land (U.S. Environmental Protection Agency, 1997). The USGS is presently contributing to these studies by studying the geochemistry of mercury in coal (Finkelman, this volume). Some of the highest mercury contents on Earth are found around mercury and gold mines. Mine wastes and soil remaining at some of these sites contain percent-level concentrations of mercury and some highly soluble compounds of mercury. More important is mercury-laden mine runoff that affects downstream ecosystems, especially contamination of water and aquatic organisms. Studies of mercury mines in the California Coast Ranges (Rytuba, this volume) show that some of these mines produce acid-water runoff that carries considerable mercury, which exceeds regulatory standards for water. Water collected downstream from mercury mines in Alaska also exceeds regulatory standards in some cases (Gray and Bailey, this volume). Furthermore, formation of the highly toxic methylmercury compound has led to mercury bioavailability, uptake, and contamination of local fish near some mercury mines (table 3). In addition to mercury entering ecosystems as a result of mercury mining, significant amounts of elemental mercury were used to recover gold during periods of historic gold mining in the United States; much of this elemental mercury was lost or was discarded to streams and rivers around these mine sites. In two of these areas, the California Sierra Nevada (Hunerlach and Alpers, this volume) and the Carson River in Nevada (Lawrence, this volume), USGS studies have shown that mercury contents are highly elevated in river-bottom sediment, water, and fish proximal to these areas. In both of these areas, significant mercury methylation has led to high mercury concentrations in fish, at levels that commonly exceed regulatory standards (table 3). Finally, the USGS is actively involved in the study of mercury emissions from volcanoes (Hinkley, this volume). This ongoing research aids in the study of mercury emissions from volcanic sources and the overall understanding of atmospheric global mercury cycling.

Table 3. Mercury and methylmercury concentrations from samples discussed in this report, comparative baselines, and regulatory standards.

[ng/L, nanograms/liter; µg/g, micrograms/gram; ng/g, nanograms/gram; -- indicates no data or not applicable]

Location	Unfiltered water		Coal, sediments, calcine, or soil		Fish (µg/g, wet weight)
	Hg (ng/L)	Methylmercury (ng/L)	Hg (µg/g)	Methylmercury (ng/g)	
Coal, USGS national database	--	--	0.003–2.9	--	--
Mercury mines, Coast Ranges, Calif.					
Below mines	1–450,000	0.01–70	0.4–220	1.1–150	0.1–1.7[1]
Mine-waste calcines	--	--	10–1,500	--	--
Mercury mines, southwestern Alaska					
Below mines	1.0–2,500	0.01–1.2	0.90–5,500	0.05–31	0.1–0.62
Mine-waste calcine and soil	--	--	3.5–46,000	0.2–41	--
Hydraulic gold mines, Sierra Nevada					
Streams below mines	40–10,400	0.01–1.1	600–26,000	0.1–0.3	0.20–1.5
Sacramento River[2]	0.03–105	0.02–1.2	0.02–0.35	0.27–2.8	--
Carson River, Nevada					
Below mines	1–28,000	0.08–7.8	0.01–100	0.56–22	0.1–16 (crayfish up to 50 µg/g)
Other Carson River studies[3-10]	3–35,400	0.4–21	0.02–1,610	0.55–29	0.1–1.4 (whole body)
Comparative baselines distal from mines					
Lake San Antonio, Calif.[4]	0.6–1.8	--	0.02–0.28	--	0.05–0.2
Truckee River, Nevada[3,9]	4.4	--	--	--	--
Pyramid Lake, Nevada[4]	1.9	--	--	--	≤0.2
Baselines streams, southwestern Alaska[11]	0.1–1.4	0.04–0.2	0.02–0.78	0.1–0.3	0.01–0.2
Uncontaminated streams, Canada[12]	--	--	0.01–0.7	--	--
Antarctica streams and lakes[13]	0.27–1.9	0.02–0.33	--	--	--
Wisconsin remote seepage lakes[14,15]	0.72–3.0	0.05–0.50	0.001–0.14	≤0.01	--
Regulatory standards					
Alaska/Nevada, State drinking water standard	2,000	--	--	--	--
EPA, acute aquatic life water standard	2,400	--	--	--	--
EPA, chronic aquatic life water standard	12	--	--	--	--
FDA, action level for fish	--	--	--	--	1.0
California, safe level for fish	--	--	--	--	0.5
EPA, fish recommendation (methylmercury)[16]	--	--	--	--	0.3

[1]Schwarzbach and others (2001). [2]Domagalski (2001). [3]Van Denburgh (1973). [4]Gill and Bruland (1990). [5]Hoffman and others (1989). [6]Gustin and others (1994). [7]Miller and others (1995). [8]Bonzongo and others (1996). [9]Wayne and others (1996). [10]Chen and others (1996). [11]Gray and others (2000). [12]Painter and others (1994). [13]Lyons and others (1999). [14]Watras and others (1994). [15]Gilmore and Reidel (1995). [16]U.S. Environmental Protection Agency (2001).

References Cited

Bonzongo, J.C., Heim, K.J., Warwick, J.J., and Lyons, W.B., 1996, Mercury levels in surface waters of the Carson River-Lahontan Reservoir system, Nevada—Influence of historic mining activities: Environmental Pollution, v. 92, p. 193–201.

Chen, Y., Bonzongo, J.C., and Miller, G.C., 1996, Levels of methylmercury and controlling factors in surface sediments of the Carson River system, Nevada: Environmental Pollution, v. 92, p. 281–287.

Domagalski, J., 2001, Mercury and methylmercury in water and sediment of the Sacramento River basin, California: Applied Geochemistry, v. 16, p. 1677–1691.

Gill, G.A., and Bruland, K.W., 1990, Mercury speciation in surface freshwater systems in California and other areas: Environmental Science and Technology, v. 24, p. 1392–1400.

Gray, J.E., Theodorakos, P.M., Bailey, E.A., and Turner, R.R., 2000, Distribution, speciation, and transport of mercury in stream sediment, stream water, and fish collected near abandoned mercury mines in southwestern Alaska, U.S.A.: Science of the Total Environment, v. 260, p. 21–33.

Gilmore, C.C., and Riedel, G.S., 1995, Measurement of Hg methylation in sediments using high specific-activity [203]Hg and ambient incubation: Water, Air, and Soil Pollution, v. 80, p. 747–756.

Gustin, M.S., Taylor, G.E., Jr., and Leonard, T.L., 1994, High levels of mercury contamination in multiple media of the Carson River drainage basin of Nevada—Implications for risk assessment: Environmental Health Perspectives, v. 102, no. 9, p. 772–778.

Hoffman, R.J., Hallock, R.J., Rowe, T.G., Lico, M.S., Burge, H.L., and Thompson, S.P., 1989, Reconnaissance investigation of water quality, bottom sediment, and biota associated with irrigation drainage in and near Stillwater Wildlife Management area, Churchill County, Nevada, 1986–87: U.S. Geological Survey Water Resources Investigations Report 89-4105, 150 p.

Lyons, W.B., Welch, K.A., and Bonzongo, J.C., 1999, Mercury in aquatic systems in Antarctica: Geophysical Research Letters, v. 26, p. 2235–2238.

Miller, J.R., Lechler, P.J., Rowland, J., Desilets, M., and Hsu, L. C., 1995, An integrated approach to the determination of the quantity, distribution, and dispersal of mercury in Lahontan Reservoir, Nevada, U.S.A.: Journal of Geochemical Exploration, v. 52, p. 45–55.

Painter, S., Cameron, E.M., Allan, R., and Rouse, J., 1994, Reconnaissance geochemistry and its environmental relevance: Journal of Geochemical Exploration, v. 51, p. 213–246.

Schwarzbach, S., Thompson, L., and Adelsbach, T., 2001, Cache Creek mercury investigation—U.S. Fish and Wildlife Service Final Report: Sacramento Fish and Wildlife Office, Environmental Contaminants Division, Office of Refuge Investigations, U.S. Fish and Wildlife Service Report FFS 1130-1F22, http://pacific.fws.gov/ecoservices/envicon/pim/reports/Sacramento/2001SacramentoCache.pdf.

U.S. Environmental Protection Agency, 1997, Mercury study report to Congress, Volume IV, An assessment of exposure to mercury in the United States: U.S. Environmental Protection Agency Report EPA-452/R-97-006.

U.S. Environmental Protection Agency, 2001, Water quality criterion for the protection of human health—Methylmercury: U.S. Environmental Protection Agency Report EPA-823-R-01-001.

Van Denburgh, A.S., 1973, Mercury in the Carson and Truckee River basins of Nevada: U.S. Geological Survey Open-File Report 73-352, 14 p.

Wayne, D.M., Warwick, J.J., Lechler, P.J., Gill, G.A., and Lyons, W.B., 1996, Mercury contamination in the Carson River, Nevada—A preliminary study of the impact of mining wastes: Water, Air, and Soil Pollution, v. 92, p. 391–408.

Watras, C.J., Bloom, N.S., Hudson, R.J.M., Gherini, S., Munson, R., Claas, S.A., Morrison, K.A., Hurley, J., Wiener, J.G., Fitzgerald, W.F., Mason, R., Vandal, G., Powell, D., Rada, R., Rislov, L., Winfrey, M., Elder, J., Krabbenhoft, D., Andren, A.W., Babiarz, C., Porcella, D.B., and Huckabee, J.W., 1994, Sources and fates of mercury and methylmercury in Wisconsin lakes, in Watras, C.J., and Huckabee, J.W., eds., Mercury pollution, integration and synthesis: Boca Raton, Fla., CRC Press, p. 153–177.